THE CONTEST PROBLEM BOOK IV

Annual High School Examinations
1973–1982
of
The Mathematical Association of America
Society of Actuaries
Mu Alpha Theta
National Council of Teachers of Mathematics
Casualty Actuarial Society

NEW MATHEMATICAL LIBRARY

PUBLISHED BY

THE MATHEMATICAL ASSOCIATION OF AMERICA

Editorial Committee

The New Mathematical Library (NML) was begun in 1961 by the School Mathematics Study Group to make available to high school students short expository books on various topics not usually covered in the high school syllabus. In a decade the NML matured into a steadily growing series of some twenty titles of interest not only to the originally intended audience, but to college students and teachers at all levels. Previously published by Random House and L. W. Singer, the NML became a publication series of the Mathematical Association of America (MAA) in 1975. Under the auspices of the MAA the NML will continue to grow and will remain dedicated to its original and expanded purposes.

THE CONTEST PROBLEM BOOK IV

Annual High School Examinations

1973–1982

compiled and with solutions by

Ralph A. Artino
The City College of New York

Anthony M. Gaglione
The U.S. Naval Academy

and

Niel Shell
The City College of New York

29

THE MATHEMATICAL ASSOCIATION
OF AMERICA

All rights reserved under International and Pan-American Copyright Conventions.
Published in Washington by the Mathematical Association of America

Library of Congress Catalog Card Number: 82-051076

Complete Set ISBN 0-88385-600-X

Vol. 29 0-88385-629-8

Manufactured in the United States of America

Contents

NEW MATHEMATICAL LIBRARY

Other titles in preparation

Preface

This volume contains the Annual High School Mathematics Examinations, given 1973 through 1982. It is a continuation of Contest Problem Books I, II, III, published as Volumes 5, 17, and 25 of the New Mathematical Library series and which contain the first twenty-three annual examinations. The Annual High School Mathematics Examinations (AHSME), it is hoped, provide challenging problems which teach, stimulate and provide enjoyment for not only the participants, but also the readers of these volumes.

All high school students are eligible to participate in the Annual High School Mathematics Examinations. In 1982, over 418,000 students in the United States, Canada, Puerto Rico, Colombia, Jamaica, Australia, Italy, England, Hungary, Ireland, Israel, Finland, Belgium and Luxembourg participated in the examination. It was administered also in many APO/FPO and other schools abroad. Each year a Summary of Results and Awards is sent to all participating high schools (in the United States and Canada). The problems are designed so that they can be solved with a knowledge of only "pre-calculus" mathematics, with emphasis on intermediate algebra and plane geometry. The subject classification at the end of the volume indicates which questions are related to which topics.

The problems on each examination become progressively more difficult. Between 1973 and 1977, the participants were given eighty minutes to complete the examination, and in subsequent years they were allowed ninety minutes. The 1973 examination consists of four parts containing 10, 10, 10 and 5 questions respectively worth 3, 4, 5, and 6 points each; to correct for random guessing, one fourth of the number of points assigned to incorrectly answered problems was deducted from the number of points assigned to correctly answered problems. The 1974 through 1977 examinations consist of 30 questions worth five points per question; one point was deducted for each question answered incorrectly. Since 1978, each examination consists of 30 questions and was scored by adding 30 points to four times the number of correct answers and then subtracting one point for each incorrect answer.

Each year since 1972, approximately one hundred of the highest scoring students on the AHSME and a number of members of previous International Mathematical Olympiad training classes have been invited to participate in the U.S.A. Mathematical Olympiad, currently a three and one half hour essay type examination consisting of five questions. Since 1974, a team of students has been selected to participate in the International Mathematical Olympiad.* An International Mathematical Olympiad training class of approximately twenty-four students receives an intensive problem solving course prior to the International Olympiad.

It is a pleasure to acknowledge the contributions of the many individuals and organizations who have made the preparation and administration of these examinations possible. We thank the members of the Committee on High School Contests and its Advisory Panel for Proposing problems and suggesting many improvements in the preliminary drafts of the examinations. We are grateful to Professor Stephen B. Maurer, who succeeded us as Committee Chairman in 1981, for his assistance in the preparation of this book. We express our appreciation to the regional examination coordinators throughout the United States and Canada who do such an excellent job of administering the examinations in their regions, and to the members of the Olympiad Subcommittee who administer all the Olympiad activities. Particular thanks are due to Professor James M. Earl, who was the chairman of the Contests Committee until his death, shortly after the 1973 examination was printed; to Professor Henry M. Cox, who was the executive director of the Contests Committee from 1973 to 1976; to Professor Walter E. Mientka, who has been the executive director of the Contests Committee since September 1976; and to Professor Samuel L. Greitzer, who has been the chairman of the Olympiad Subcommittee since the inception of the subcommittee. We express appreciation to our sponsors, the Mathematical Association of America, the Society of Actuaries, Mu Alpha Theta, the National Council of Teachers of Mathematics, and the Casualty Actuarial Society for their financial support and guidance; we thank the City College of New York, the University of Nebraska, Swarthmore College and Metropolitan Life Insurance Company for the support they have provided present and past chairmen and executive directors; and we thank L. G. Balfour Company, W. H. Freeman and Company, Kuhn Associates, National Semiconductor, Pickett, Inc., MAA, Mu

---

*The International Olympiads from 1959 to 1977 have been published in volume 27 of the New Mathematical Library series.

Alpha Theta, NCTM and Random House for donating awards to high-scoring individuals and schools on the AHSME.

The members of the Committee on High School Contests are particularly pleased to acknowledge financial support for the U.S.A. Mathematical Olympiad and the participation of the U.S. team in the International Mathematical Olympiad. We express our gratitude to the International Business Machines Corporation for an annual grant to sponsor an awards ceremony in honor of the winners of the U.S.A. Mathematical Olympiad; we thank the hosts of training sessions: Rutgers University, United States Military Academy and United States Naval Academy; we gratefully acknowledge financial support of the training sessions and travel to the International Mathematical Olympiad from the following: Army Research Office, Johnson and Johnson Foundation, Office of Naval Research, Minnesota Mining & Manufacturing Corporation, National Science Foundation, Spencer Foundation, Standard Oil Company of California and Xerox Corporation.

A few minor changes in the statements of problems have been made in this collection for the sake of greater clarity.

Ralph A. Artino

Anthony M. Gaglione

Niel Shell

Editors' Preface

The editors of the New Mathematical Library, wishing to encourage significant problem solving at an elementary level, have published a variety of problem collections. In addition to the Annual High School Mathematics Examinations (NML vols. 5, 17, 25 and 29) described in detail in the Preface on the preceding pages, the NML series contains translations of the Hungarian Eötvös Competitions through 1928 (NML vols. 11 and 12) and the International Mathematical Olympiads (NML vol. 27). Both are essay type competitions containing only a few questions which often require ingenious solutions.

The present volume is a sequel to NML vol. 25 published at the request of the many readers who enjoyed the previous Contest Problem Books.

The Mathematical Association of America, publisher of the NML series, is concerned primarily with mathematics at the undergraduate level in colleges and universities. It conducts the annual Putnam Competitions for undergraduate students. All three journals of the MAA, the *American Mathematical Monthly, Mathematics Magazine* and the *Two Year College Mathematics Journal*, have sections devoted to problems and their solutions.

The editors of the New Mathematical Library are pleased to acknowledge the essential contributions of R. A. Artino, A. M. Gaglione and N. Shell, the three men who compiled and wrote solutions for the problems in the present volume. The hard work of Stephen B. Maurer, current chairman of the MAA Committee on High School Contests, and that of other Committee members in the final editing of this collection is greatly appreciated.

We suggest that readers attempt their own solutions before looking at the ones offered. Their solutions may be quite different from, but just as good or better than, those published here.

People taking the AHSME are told to avoid random guessing, since there is a penalty for incorrect answers; however, if a participant can use his mathematical knowledge to eliminate all but one of the listed choices, he will improve his score. A few examples of this kind of elimination are indicated in some of the Notes appended to solutions.

Basil Gordon

Anneli Lax

1982

List of Symbols

Symbol	Name and/or Meaning
$\{x: \ \}$	set of all x such that; e.g. $\{x : x$ is a positive integer less than 4$\}$ is the set with members 1, 2, 3
$\subset$	contained in; $A \subset B$ means each member of A is in B
$\supset$	contains: $A \supset B$ means $B \subset A$
$\neq$	not equal to
$f(x)$	function f of the variable x
$f(a)$	the value that f assigns to the constant a
$\equiv$	identically equal; e.g. $f(x) \equiv 1$ means $f(x) = 1$ for all values of the variable x
$\equiv$	in number theory, for integers $a, b, m,$ $\quad a \equiv b(\text{mod } m)$ is read "a is congruent to b mod m" and means that $a - b$ is divisible by m.
$<$	less than
$\leqslant$	less than or equal to
$>$	greater than
$\geqslant$	greater than or equal to
$\approx$	approximately equal to
$\|x\|$	absolute value; $\|x\| = \left\{ \begin{array}{l} x \text{ if } x \geqslant 0 \\ -x \text{ if } x < 0 \end{array} \right\}$
$n!$	n factorial; $n! = 1 \cdot 2 \cdot \ldots \cdot n$
$\binom{n}{k}$	combinations of n things taken k at a time; $\binom{n}{k} = \dfrac{n!}{k!(n-k)!}$
$\displaystyle\sum_{i=1}^{p}$	summation sign; $\displaystyle\sum_{i=1}^{n} a_i$ means $a_1 + a_2 + \ldots + a_n$
$\displaystyle\sum_{i=1}^{\infty}$	infinite sum; $\displaystyle\sum_{i=1}^{\infty} a_i = a_1 + a_2 + \ldots$
$(abcd)_n$	base n representation; $an^3 + bn^2 + cn + d$
Δa_n	first difference; for a sequence $a_1, a_2, \ldots,$ Δa_n means $a_{n+1} - a_n$
$\Delta^k a_n$	kth difference; $\Delta^1 a_n = \Delta a_n$, $\Delta^{k+1} a_n = \Delta (\Delta^k a_n)$ for $k > 1$
$[x]$	the largest integer not bigger than x

Symbol	Name and/or Meaning
$\begin{vmatrix} a & b \\ c & d \end{vmatrix}$	determinant of the matrix $\begin{pmatrix} a & b \\ c & d \end{pmatrix}$, equal to $ad - bc$
i	imaginary unit in the set of complex numbers, satisfying $i^2 = -1$; also used as a constant or as a variable (or index) taking integer values (e.g. in $\sum_{i=1}^{5} a_i = a_1 + a_2 + \cdots + a_5$)
$\bar{z}$	complex conjugate of z; if $z = a + ib$ with a, b real, then $\bar{z} = a - bi$.
AB	either line segment connecting points A and B or its length
$\overset{\frown}{AB}$	the minor circular arc with endpoints A and B
$\angle ABC$	either angle ABC or its measure
$\perp$	is perpendicular to
$\parallel$	is parallel to
$\sim$	is similar to
$\cong$	is congruent to
$\square$	parallelogram

I

Problems

1973 Examination

Part 1

1. A chord which is the perpendicular bisector of a radius of length 12 in a circle, has length

   (A) $3\sqrt{3}$ (B) 27   (C) $6\sqrt{3}$ (D) $12\sqrt{3}$ (E) none of these

2. One thousand unit cubes are fastened together to form a large cube with edge length 10 units; this is painted and then separated into the original cubes. The number of these unit cubes which have at least one face painted is

   (A) 600   (B) 520   (C) 488   (D) 480   (E) 400

3. The stronger Goldbach conjecture states that any even integer greater than 7 can be written as the sum of two different prime numbers.[†] For such representations of the even number 126, the largest possible difference between the two primes is

   (A) 112   (B) 100   (C) 92   (D) 88   (E) 80

[†] The regular Goldbach conjecture states that any even integer greater than 3 is expressible as a sum of two primes. Neither this conjecture nor the stronger version has been settled.

4. Two congruent $30° - 60° - 90°$ triangles are placed so that they overlap partly and their hypotenuses coincide. If the hypotenuse of each triangle is 12, the area common to both triangles is

(A) $6\sqrt{3}$ (B) $8\sqrt{3}$ (C) $9\sqrt{3}$ (D) $12\sqrt{3}$ (E) 24

5. Of the following five statements, I to V, about the binary operation of averaging (arithmetic mean),

   I. Averaging is associative
   II. Averaging is commutative
   III. Averaging distributes over addition
   IV. Addition distributes over averaging
   V. Averaging has an identity element

those which are always true are

(A) All   (B) I and II only   (C) II and III only
(D) II and IV only   (E) II and V only

6. If 554 is the base b representation of the square of the number whose base b representation is 24, then b, when written in base 10, equals

(A) 6   (B) 8   (C) 12   (D) 14   (E) 16

7. The sum of all the integers between 50 and 350 which end in 1 is

(A) 5880   (B) 5539   (C) 5208   (D) 4877   (E) 4566

8. If 1 pint of paint is needed to paint a statue 6 ft. high, then the number of pints it will take to paint (to the same thickness) 540 statues similar to the original but only 1 ft. high, is

(A) 90   (B) 72   (C) 45   (D) 30   (E) 15

9. In $\triangle ABC$ with right angle at C, altitude CH and median CM trisect the right angle. If the area of $\triangle CHM$ is K, then the area of $\triangle ABC$ is

(A) $6K$ (B) $4\sqrt{3}\,K$ (C) $3\sqrt{3}\,K$ (D) $3K$ (E) $4K$

10. If n is a real number, then the simultaneous system to the right has no solution if and only if n is equal to

$$\begin{cases} nx + y = 1 \\ ny + z = 1 \\ x + nz = 1 \end{cases}$$

(A) -1 (B) 0   (C) 1
(D) 0 or 1   (E) $\frac{1}{2}$

Part 2

11. A circle with a circumscribed and an inscribed square centered at the origin O of a rectangular coordinate system with positive x and y axes OX and OY is shown in each figure I to IV below.

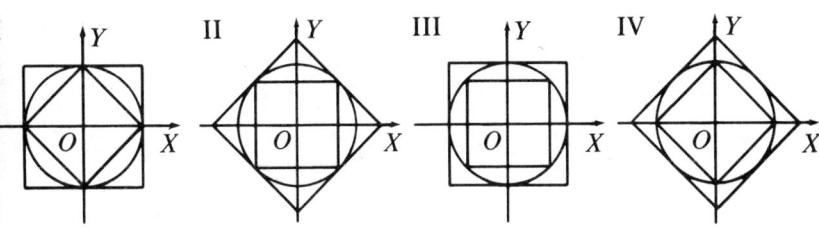

The inequalities

$$|x| + |y| \leqslant \sqrt{2(x^2 + y^2)} \leqslant 2\,\text{Max}(|x|, |y|)$$

are represented geometrically[†] by the figure numbered

(A) I   (B) II   (C) III   (D) IV   (E) none of these

12. The average (arithmetic mean) age of a group consisting of doctors and lawyers is 40. If the doctors average 35 and the lawyers 50 years old, then the ratio of the number of doctors to the number of lawyers is

(A) 3:2   (B) 3:1   (C) 2:3   (D) 2:1   (E) 1:2

13. The fraction $\dfrac{2(\sqrt{2} + \sqrt{6})}{3\sqrt{2 + \sqrt{3}}}$ is equal to

(A) $\dfrac{2\sqrt{2}}{3}$ (B) 1   (C) $\dfrac{2\sqrt{3}}{3}$ (D) $\dfrac{4}{3}$ (E) $\dfrac{16}{9}$

[†]An inequality of the form $f(x, y) \leqslant g(x, y)$, for all x and y is represented geometrically by a figure showing the containment

{The set of points (x, y) such that $g(x, y) \leqslant a$}

$\subset$ {The set of points (x, y) such that $f(x, y) \leqslant a$}

for a typical real number a.

14. Each valve A, B, and C, when open, releases water into a tank at its own constant rate. With all three valves open, the tank fills in 1 hour, with only valves A and C open it takes 1.5 hour, and with only valves B and C open it takes 2 hours. The number of hours required with only valves A and B open is

(A) 1.1   (B) 1.15   (C) 1.2   (D) 1.25   (E) 1.75

15. A sector with acute central angle θ is cut from a circle of radius 6. The radius of the circle circumscribed about the sector is

(A) $3\cos\theta$ (B) $3\sec\theta$ (C) $3\cos\frac{1}{2}\theta$ (D) $3\sec\frac{1}{2}\theta$ (E) 3

16. If the sum of all the angles except one of a convex polygon is $2190°$, then the number of sides of the polygon must be

(A) 13   (B) 15   (C) 17   (D) 19   (E) 21

17. If θ is an acute angle and $\sin\frac{1}{2}\theta = \sqrt{\dfrac{x-1}{2x}}$, then $\tan\theta$ equals

(A) x (B) $\dfrac{1}{x}$ (C) $\dfrac{\sqrt{x-1}}{x+1}$ (D) $\dfrac{\sqrt{x^2-1}}{x}$ (E) $\sqrt{x^2-1}$

18. If $p \geqslant 5$ is a prime number, then 24 divides $p^2 - 1$ without remainder

(A) never   (B) sometimes only   (C) always
(D) only if $p = 5$ (E) none of these

19. Define $n_a!$ for n and a positive to be

$$n_a! = n(n-a)(n-2a)(n-3a)\ldots(n-ka),$$

where k is the greatest integer for which $n > ka$.
Then the quotient $72_8!/18_2!$ is equal to

(A) 4^5 (B) 4^6 (C) 4^8 (D) 4^9 (E) 4^{12}

20. A cowboy is 4 miles south of a stream which flows due east. He is also 8 miles west and 7 miles north of his cabin. He wishes to water his horse at the stream and return home. The shortest distance (in miles) he can travel and accomplish this is

(A) $4 + \sqrt{185}$ (B) 16   (C) 17   (D) 18   (E) $\sqrt{32} + \sqrt{137}$

Part 3

21. The number of sets of two or more consecutive positive integers whose sum is 100 is

    (A) 1   (B) 2   (C) 3   (D) 4   (E) 5

22. The set of all real solutions of the inequality
$$|x - 1| + |x + 2| < 5$$
    is

    (A) $\{x: -3 < x < 2\}$ (B) $\{x: -1 < x < 2\}$
    (C) $\{x: -2 < x < 1\}$ (D) $\{x: -\frac{3}{2} < x < \frac{7}{2}\}$ (E) $\varnothing$ (empty)

23. There are two cards; one is red on both sides and the other is red on one side and blue on the other. The cards have the same probability ($\frac{1}{2}$) of being chosen, and one is chosen and placed on the table. If the upper side of the card on the table is red, then the probability that the under-side is also red is

    (A) $\frac{1}{4}$ (B) $\frac{1}{3}$ (C) $\frac{1}{2}$ (D) $\frac{2}{3}$ (E) $\frac{3}{4}$

24. The check for a luncheon of 3 sandwiches, 7 cups of coffee and one piece of pie came to \$3.15. The check for a luncheon consisting of 4 sandwiches, 10 cups of coffee and one piece of pie came to \$4.20 at the same place. The cost of a luncheon consisting of one sandwich, one cup of coffee and one piece of pie at the same place will come to

    (A) \$1.70   (B) \$1.65   (C) \$1.20   (D) \$1.05   (E) \$.95

25. A circular grass plot 12 feet in diameter is cut by a straight gravel path 3 feet wide, one edge of which passes through the center of the plot. The number of square feet in the remaining grass area is

    (A) $36\pi - 34$ (B) $30\pi - 15$ (C) $36\pi - 33$
    (D) $35\pi - 9\sqrt{3}$ (E) $30\pi - 9\sqrt{3}$

26. The number of terms in an A.P. (Arithmetic Progression) is even. The sums of the odd- and even-numbered terms are 24 and 30 respectively. If the last term exceeds the first by 10.5, the number of terms in the A.P. is

    (A) 20   (B) 18   (C) 12   (D) 10   (E) 8

27. Cars A and B travel the same distance. Car A travels half that *distance* at u miles per hour and half at v miles per hour. Car B travels half the *time* at u miles per hour and half at v miles per hour. The average speed of Car A is x miles per hour and that of Car B is y miles per hour: Then we always have

   (A) $x \leqslant y$ (B) $x \geqslant y$ (C) $x = y$ (D) $x < y$ (E) $x > y$

28. If a, b, and c are in geometric progression (G.P.) with $1 < a < b < c$ and $n > 1$ is an integer, then $\log_a n, \log_b n, \log_c n$ form a sequence

   (A) which is a G.P.
   (B) which is an arithmetic progression (A.P.)
   (C) in which the reciprocals of the terms form an A.P.
   (D) in which the second and third terms are the nth powers of the first and second respectively
   (E) none of these

29. Two boys start moving from the same point A on a circular track but in opposite directions. Their speeds are 5 ft. per sec. and 9 ft. per sec. If they start at the same time and finish when they first meet at the point A again, then the number of times they meet, excluding the start and finish, is

   (A) 13   (B) 25   (C) 44   (D) infinity   (E) none of these

30. Let $[t]$ denote the greatest integer $\leqslant t$ where $t \geqslant 0$ and $S = \{(x, y): (x - T)^2 + y^2 \leqslant T^2$ where $T = t - [t]\}$. Then we have

   (A) the point $(0, 0)$ does not belong to S for any t
   (B) $0 \leqslant$ Area $S \leqslant \pi$ for all t
   (C) S is contained in the first quadrant for all $t \geqslant 5$
   (D) the center of S for any t is on the line $y = x$
   (E) none of the other statements is true

Part 4

31. In the following equation, each of the letters represents uniquely a different digit in base ten:

$$(YE) \cdot (ME) = TTT$$

   The sum $E + M + T + Y$ equals

   (A) 19   (B) 20   (C) 21   (D) 22   (E) 24

32. The volume of a pyramid whose base is an equilateral triangle of side length 6 and whose other edges are each of length $\sqrt{15}$ is

   (A) 9   (B) 9/2   (C) 27/2   (D) $\dfrac{9\sqrt{3}}{2}$ (E) none of these

33. When one ounce of water is added to a mixture of acid and water, the new mixture is 20% acid. When one ounce of acid is added to the new mixture, the result is $33\frac{1}{3}\%$ acid. The percentage of acid in the original mixture is

   (A) 22%   (B) 24%   (C) 25%   (D) 30%   (E) $33\frac{1}{3}\%$

34. A plane flew straight against a wind between two towns in 84 minutes and returned with that wind in 9 minutes less than it would take in still air. The number of minutes (2 answers) for the return trip was

   (A) 54 or 18   (B) 60 or 15   (C) 63 or 12   (D) 72 or 36
   (E) 75 or 20

35. In the unit circle shown in the figure to the right, chords PQ and MN are parallel to the unit radius OR of the circle with center at O. Chords MP, PQ and NR are each s units long and chord MN is d units long. Of the three equations

   I. $d - s = 1$,   II. $ds = 1$,
   III. $d^2 - s^2 = \sqrt{5}$

   those which are necessarily true are

   (A) I only   (B) II only   (C) III only
   (D) I and II only   (E) I, II, and III

1974 Examination

1. If $x \neq 0$ or 4 and $y \neq 0$ or 6, then $\dfrac{2}{x} + \dfrac{3}{y} = \dfrac{1}{2}$ is equivalent to

   (A) $4x + 3y = xy$ (B) $y = \dfrac{4x}{6 - y}$ (C) $\dfrac{x}{2} + \dfrac{y}{3} = 2$

   (D) $\dfrac{4y}{y - 6} = x$ (E) none of these

2. Let x_1 and x_2 be such that $x_1 \neq x_2$ and $3x_i^2 - hx_i = b$, $i = 1, 2$. Then $x_1 + x_2$ equals

   (A) $-\dfrac{h}{3}$ (B) $\dfrac{h}{3}$ (C) $\dfrac{b}{3}$ (D) $2b$ (E) $-\dfrac{b}{3}$

3. The coefficient of x^7 in the polynomial expansion of

   $$(1 + 2x - x^2)^4$$

   is

   (A) -8 (B) 12   (C) 6   (D) -12 (E) none of these

4. What is the remainder when $x^{51} + 51$ is divided by $x + 1$?
   (A) 0   (B) 1   (C) 49   (D) 50   (E) 51

5. Given a quadrilateral $ABCD$ inscribed in a circle with side AB extended beyond B to point E, if $\angle BAD = 92°$ and $\angle ADC = 68°$, find $\angle EBC$.

   (A) $66°$ (B) $68°$ (C) $70°$ (D) $88°$ (E) $92°$

6. For positive real numbers x and y define $x * y = \dfrac{x \cdot y}{x + y}$; then

   (A) " $*$ " is commutative but not associative
   (B) " $*$ " is associative but not commutative
   (C) " $*$ " is neither commutative nor associative
   (D) " $*$ " is commutative and associative
   (E) none of these

7. A town's population increased by 1,200 people, and then this new population decreased by 11%. The town now had 32 less people than it did before the 1,200 increase. What is the original population?

(A) 1,200   (B) 11,200   (C) 9,968   (D) 10,000
(E) none of these

8. What is the smallest prime number dividing the sum $3^{11} + 5^{13}$?

(A) 2   (B) 3   (C) 5   (D) $3^{11} + 5^{13}$ (E) none of these

9. The integers greater than one are arranged in five columns as follows:

$$
\begin{array}{ccccc}
 & 2 & 3 & 4 & 5 \\
9 & 8 & 7 & 6 & \\
 & 10 & 11 & 12 & 13 \\
17 & 16 & 15 & 14 & \\
 & \cdot & \cdot & \cdot & \cdot
\end{array}
$$

(Four consecutive integers appear in each row; in the first, third and other odd numbered rows, the integers appear in the last four columns and increase from left to right; in the second, fourth and other even numbered rows, the integers appear in the first four columns and increase from right to left.)

In which column will the number 1,000 fall?

(A) first   (B) second   (C) third   (D) fourth   (E) fifth

10. What is the smallest integral value of k such that

$$2x(kx - 4) - x^2 + 6 = 0$$

has no real roots?

(A) -1 (B) 2   (C) 3   (D) 4   (E) 5

11. If (a, b) and (c, d) are two points on the line whose equation is $y = mx + k$, then the distance between (a, b) and (c, d), in terms of a, c and m, is

(A) $|a - c|\sqrt{1 + m^2}$ (B) $|a + c|\sqrt{1 + m^2}$ (C) $\dfrac{|a - c|}{\sqrt{1 + m^2}}$

(D) $|a - c|(1 + m^2)$ (E) $|a - c||m|$

12. If $g(x) = 1 - x^2$ and $f(g(x)) = \dfrac{1 - x^2}{x^2}$ when $x \neq 0$, then $f(1/2)$ equals

(A) $3/4$ (B) 1 (C) 3 (D) $\sqrt{2}/2$ (E) $\sqrt{2}$

13. Which of the following is equivalent to "If P is true then Q is false."?

(A) "P is true or Q is false."
(B) "If Q is false then P is true."
(C) "If P is false then Q is true."
(D) "If Q is true then P is false."
(E) "If Q is true then P is true."

14. Which statement is correct?

(A) If $x < 0$, then $x^2 > x$.   (B) If $x^2 > 0$, then $x > 0$.
(C) If $x^2 > x$, then $x > 0$.   (D) If $x^2 > x$, then $x < 0$.
(E) If $x < 1$, then $x^2 < x$.

15. If $x < -2$ then $|1 - |1 + x||$ equals

(A) $2 + x$ (B) $-2 - x$ (C) x (D) $-x$ (E) -2

16. A circle of radius r is inscribed in a right isosceles triangle, and a circle of radius R is circumscribed about the triangle. Then R/r equals

(A) $1 + \sqrt{2}$ (B) $\dfrac{2 + \sqrt{2}}{2}$ (C) $\dfrac{\sqrt{2} - 1}{2}$

(D) $\dfrac{1 + \sqrt{2}}{2}$ (E) $2(2 - \sqrt{2})$

17. If $i^2 = -1$, then $(1 + i)^{20} - (1 - i)^{20}$ equals

(A) -1024 (B) $-1024i$ (C) 0 (D) 1024 (E) $1024i$

18. If $\log_8 3 = p$ and $\log_3 5 = q$, then, in terms of p and q, $\log_{10} 5$ equals

(A) pq (B) $\dfrac{3p + q}{5}$ (C) $\dfrac{1 + 3pq}{p + q}$ (D) $\dfrac{3pq}{1 + 3pq}$

(E) $p^2 + q^2$

19. In the adjoining figure $ABCD$ is a square and CMN is an equilateral triangle. If the area of $ABCD$ is one square inch, then the area of CMN in square inches is

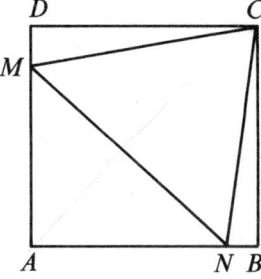

(A) $2\sqrt{3} - 3$ (B) $1 - \sqrt{3}/3$

(C) $\sqrt{3}/4$ (D) $\sqrt{2}/3$

(E) $4 - 2\sqrt{3}$

20. Let

$$T = \frac{1}{3 - \sqrt{8}} - \frac{1}{\sqrt{8} - \sqrt{7}} + \frac{1}{\sqrt{7} - \sqrt{6}} - \frac{1}{\sqrt{6} - \sqrt{5}} + \frac{1}{\sqrt{5} - 2};$$

then

(A) $T < 1$ (B) $T = 1$ (C) $1 < T < 2$ (D) $T > 2$

(E) $T = \dfrac{1}{(3 - \sqrt{8})(\sqrt{8} - \sqrt{7})(\sqrt{7} - \sqrt{6})(\sqrt{6} - \sqrt{5})(\sqrt{5} - 2)}$

21. In a geometric series of positive terms the difference between the fifth and fourth terms is 576, and the difference between the second and first terms is 9. What is the sum of the first five terms of this series?

(A) 1061 (B) 1023 (C) 1024 (D) 768 (E) none of these

22. The minimum value of $\sin\dfrac{A}{2} - \sqrt{3}\cos\dfrac{A}{2}$ is attained when A is

(A) $-180°$ (B) $60°$ (C) $120°$ (D) $0°$ (E) none of these

23. In the adjoining figure TP and $T'Q$ are parallel tangents to a circle of radius r, with T and T' the points of tangency. $PT''Q$ is a third tangent with T'' as point of tangency. If $TP = 4$ and $T'Q = 9$ then r is

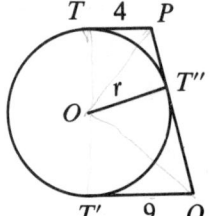

(A) $25/6$ (B) 6 (C) $25/4$
(D) a number other than $25/6$, 6, $25/4$
(E) not determinable from the given information

24. A fair die is rolled six times. The probability of rolling at least a five at least five times is

(A) $13/729$ (B) $12/729$ (C) $2/729$ (D) $3/729$
(E) none of these

25. In parallelogram $ABCD$ of the accompanying diagram, line DP is drawn bisecting BC at N and meeting AB (extended) at P. From vertex C, line CQ is drawn bisecting side AD at M and meeting AB (extended) at Q. Lines DP and CQ meet at O. If the area of parallelogram $ABCD$ is k, then the area of triangle QPO is equal to

(A) k (B) $6k/5$

(C) $9k/8$ (D) $5k/4$

(E) $2k$

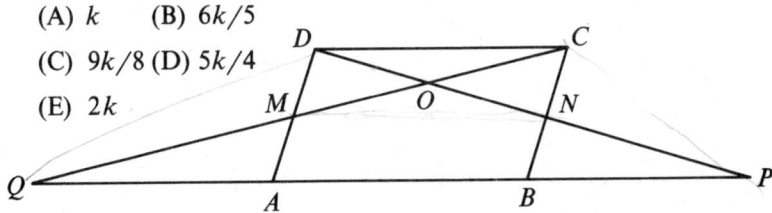

26. The number of distinct positive integral divisors of $(30)^4$ excluding 1 and $(30)^4$ is

(A) 100 (B) 125 (C) 123 (D) 30 (E) none of these

27. If $f(x) = 3x + 2$ for all real x, then the statement:

"$|f(x) + 4| < a$ whenever $|x + 2| < b$ and $a > 0$ and $b > 0$" is true when

(A) $b \leqslant a/3$ (B) $b > a/3$ (C) $a \leqslant b/3$ (D) $a > b/3$

(E) The statement is never true.

28. Which of the following is satisfied by all numbers x of the form

$$x = \frac{a_1}{3} + \frac{a_2}{3^2} + \cdots + \frac{a_{25}}{3^{25}},$$

where a_1 is 0 or 2, a_2 is 0 or 2,..., a_{25} is 0 or 2?

(A) $0 \leqslant x < 1/3$ (B) $1/3 \leqslant x < 2/3$ (C) $2/3 \leqslant x < 1$

(D) $0 \leqslant x < 1/3$ or $2/3 \leqslant x < 1$ (E) $1/2 \leqslant x \leqslant 3/4$

29. For $p = 1, 2, \ldots, 10$ let S_p be the sum of the first 40 terms of the arithmetic progression whose first term is p and whose common difference is $2p - 1$; then $S_1 + S_2 + \cdots + S_{10}$ is

(A) 80,000   (B) 80,200   (C) 80,400   (D) 80,600
(E) 80,800

30. A line segment is divided so that the lesser part is to the greater part as the greater part is to the whole. If R is the ratio of the lesser part to the greater part, then the value of

$$R^{[R^{(R^2+R^{-1})}+R^{-1}]} + R^{-1}$$

is

(A) 2   (B) $2R$ (C) R^{-1} (D) $2 + R^{-1}$ (E) $2 + R$

1975 Examination

1. The value of $\dfrac{1}{2 - \dfrac{1}{2 - \dfrac{1}{2 - \frac{1}{2}}}}$ is

   (A) 3/4   (B) 4/5   (C) 5/6   (D) 6/7   (E) 6/5

2. For which real values of m are the simultaneous equations

$$y = mx + 3$$
$$y = (2m - 1)x + 4$$

   satisfied by at least one pair of real numbers (x, y)?

   (A) all m (B) all $m \neq 0$ (C) all $m \neq 1/2$ (D) all $m \neq 1$
   (E) no values of m

3. Which of the following inequalities are satisfied for all real numbers a, b, c, x, y, z which satisfy the conditions $x < a, y < b$, and $z < c$?

   I. $xy + yz + zx < ab + bc + ca$
   II. $x^2 + y^2 + z^2 < a^2 + b^2 + c^2$
   III. $xyz < abc$

   (A) None are satisfied.   (B) I only   (C) II only
   (D) III only   (E) All are satisfied.

4. If the side of one square is the diagonal of a second square, what is the ratio of the area of the first square to the area of the second?

   (A) 2   (B) $\sqrt{2}$ (C) 1/2   (D) $2\sqrt{2}$ (E) 4

5. The polynomial $(x + y)^9$ is expanded in decreasing powers of x. The second and third terms have equal values when evaluated at $x = p$ and $y = q$, where p and q are positive numbers whose sum is one. What is the value of p?

   (A) 1/5   (B) 4/5   (C) 1/4   (D) 3/4   (E) 8/9

6. The sum of the first eighty positive odd integers subtracted from the sum of the first eighty positive even integers is

   (A) 0   (B) 20   (C) 40   (D) 60   (E) 80

7. For which non-zero real numbers x is $\dfrac{|x - |x||}{x}$ a positive integer?

(A) for negative x only   (B) for positive x only
(C) only for x an even integer
(D) for all non-zero real numbers x
(E) for no non-zero real numbers x

8. If the statement "All shirts in this store are on sale." is false, then which of the following statements must be true?

I. All shirts in this store are at non-sale prices.[†]
II. There is some shirt in this store not on sale.
III. No shirt in this store is on sale.
IV. Not all shirts in this store are on sale.

(A) II only   (B) IV only   (C) I and III only
(D) II and IV only   (E) I, II and IV only

9. Let $a_1, a_2, \ldots$ and $b_1, b_2, \ldots$ be arithmetic progressions such that $a_1 = 25$, $b_1 = 75$ and $a_{100} + b_{100} = 100$. Find the sum of the first one hundred terms of the progression $a_1 + b_1, a_2 + b_2, \ldots$.

(A) 0   (B) 100   (C) 10,000   (D) 505,000
(E) not enough information given to solve the problem

10. The sum of the digits in base ten of $(10^{4n^2+8} + 1)^2$, where n is a positive integer, is

(A) 4   (B) $4n$ (C) $2 + 2n$ (D) $4n^2$ (E) $n^2 + n + 2$

11. Let P be an interior point of circle K other than the center of K. Form all chords of K which pass through P, and determine their midpoints. The locus of these midpoints is

(A) a circle with one point deleted
(B) a circle if the distance from P to the center of K is less than one half the radius of K; otherwise a circular arc of less than $360°$
(C) a semicircle with one point deleted
(D) a semicircle   (E) a circle

[†]Originally, statement I read: All shirts in this store are not on sale.

12. If $a \neq b$, $a^3 - b^3 = 19x^3$ and $a - b = x$, which of the following conclusions is correct?

(A) $a = 3x$ (B) $a = 3x$ or $a = -2x$
(C) $a = -3x$ or $a = 2x$ (D) $a = 3x$ or $a = 2x$ (E) $a = 2x$

13. The equation $x^6 - 3x^5 - 6x^3 - x + 8 = 0$ has

(A) no real roots
(B) exactly two distinct negative roots
(C) exactly one negative root
(D) no negative roots, but at least one positive root
(E) none of these

14. If the *whatsis* is *so* when the *whosis* is *is* and the *so* and *so* is *is · so*, what is the *whosis · whatsis* when the *whosis* is *so*, the *so* and *so* is *so · so*, and the *is* is two (*whatsis, whosis, is* and *so* are variables taking positive values)?

(A) *whosis · is · so*   (B) *whosis*   (C) *is*   (D) *so*
(E) *so* and *so*

15. In the sequence of numbers $1, 3, 2, \ldots$ each term after the first two is equal to the term preceding it minus the term preceding that. The sum of the first one hundred terms of the sequence is

(A) 5   (B) 4   (C) 2   (D) 1   (E) -1

16. If the first term of an infinite geometric series is a positive integer, the common ratio is the reciprocal of a positive integer, and the sum of the series is 3, then the sum of the first two terms of the series is

(A) $1/3$ (B) $2/3$ (C) $8/3$ (D) 2   (E) $9/2$

17. A man can commute either by train or by bus. If he goes to work on the train in the morning, he comes home on the bus in the afternoon; and if he comes home in the afternoon on the train, he took the bus in the morning. During a total of x working days, the man took the bus to work in the morning 8 times, came home by bus in the afternoon 15 times, and commuted by train (either morning or afternoon) 9 times. Find x.

(A) 19   (B) 18   (C) 17   (D) 16
(E) not enough information given to solve the problem

18. A positive integer N with three digits in its base ten representation is chosen at random, with each three digit number having an equal chance of being chosen. The probability that $\log_2 N$ is an integer is

    (A) 0   (B) 3/899   (C) 1/225   (D) 1/300   (E) 1/450

19. Which positive numbers x satisfy the equation $(\log_3 x)(\log_x 5) = \log_3 5$?

    (A) 3 and 5 only   (B) 3, 5 and 15 only
    (C) only numbers of the form $5^n \cdot 3^m$, where n and m are positive integers
    (D) all positive $x \neq 1$ (E) none of these

20. In the adjoining figure triangle ABC is such that $AB = 4$ and $AC = 8$. If M is the midpoint of BC and $AM = 3$, what is the length of BC?

    (A) $2\sqrt{26}$ (B) $2\sqrt{31}$
    (C) 9   (D) $4 + 2\sqrt{13}$
    (E) not enough information
        given to solve the problem

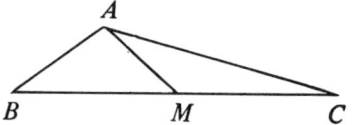

21. Suppose $f(x)$ is defined for all real numbers x; $f(x) > 0$ for all x; and $f(a)f(b) = f(a + b)$ for all a and b. Which of the following statements are true?

    I. $f(0) = 1$ II. $f(-a) = 1/f(a)$ for all a

    III. $f(a) = \sqrt[3]{f(3a)}$ for all a IV. $f(b) > f(a)$ if $b > a$

    (A) III and IV only   (B) I, III and IV only
    (C) I, II and IV only (D) I, II and III only (E) All are true.

22. If p and q are primes and $x^2 - px + q = 0$ has distinct positive integral roots, then which of the following statements are true?

    I. The difference of the roots is odd.
    II. At least one root is prime.
    III. $p^2 - q$ is prime.
    IV. $p + q$ is prime.

    (A) I only   (B) II only   (C) II and III only
    (D) I, II and IV only   (E) All are true.

23. In the adjoining figure AB and BC are adjacent sides of square $ABCD$; M is the midpoint of AB; N is the midpoint of BC; and AN and CM intersect at O. The ratio of the area of $AOCD$ to the area of $ABCD$ is

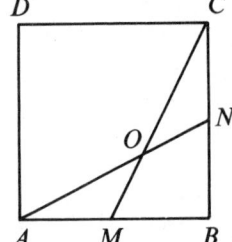

(A) 5/6   (B) 3/4   (C) 2/3
(D) $\sqrt{3}/2$ (E) $(\sqrt{3} - 1)/2$

24. In triangle ABC, $\angle C = \theta$ and $\angle B = 2\theta$, where $0° < \theta < 60°$. The circle with center A and radius AB intersects AC at D and intersects BC, extended if necessary, at B and at E (E may coincide with B). Then $EC = AD$

(A) for no values of θ (B) only if $\theta = 45°$
(C) only if $0° < \theta \leqslant 45°$ (D) only if $45° \leqslant \theta < 60°$
(E) for all θ such that $0° < \theta < 60°$

25. A woman, her brother, her son and her daughter are chess players (all relations by birth). The worst player's twin (who is one of the four players) and the best player are of opposite sex. The worst player and the best player are the same age. Who is the worst player?

(A) the woman   (B) her son
(C) her brother   (D) her daughter
(E) No solution is consistent with the given information.

26. In acute triangle ABC the bisector of $\angle A$ meets side BC at D. The circle with center B and radius BD intersects side AB at M; and the circle with center C and radius CD intersects side AC at N. Then it is always true that

(A) $\angle CND + \angle BMD - \angle DAC = 120°$
(B) $AMDN$ is a trapezoid
(C) BC is parallel to MN
(D) $AM - AN = \dfrac{3(DB - DC)}{2}$
(E) $AB - AC = \dfrac{3(DB - DC)}{2}$

27. If p, q and r are distinct roots of $x^3 - x^2 + x - 2 = 0$, then $p^3 + q^3 + r^3$ equals

(A) -1 (B) 1   (C) 3   (D) 5   (E) none of these

28. In triangle ABC shown in the adjoining figure, M is the midpoint of side BC, $AB = 12$ and $AC = 16$. Points E and F are taken on AC and AB, respectively, and lines EF and AM intersect at G. If $AE = 2AF$ then EG/GF equals

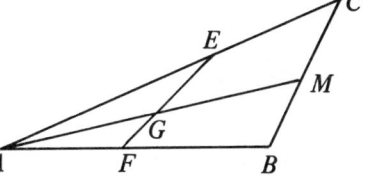

(A) $3/2$ (B) $4/3$
(C) $5/4$ (D) $6/5$
(E) not enough information
    given to solve the problem

29. What is the smallest integer larger than $(\sqrt{3} + \sqrt{2})^6$?

(A) 972   (B) 971   (C) 970   (D) 969   (E) 968

30. Let $x = \cos 36° - \cos 72°$. Then x equals

(A) $1/3$ (B) $1/2$ (C) $3 - \sqrt{6}$ (D) $2\sqrt{3} - 3$
(E) none of these

1976 Examination

1. If one minus the reciprocal of $(1 - x)$ equals the reciprocal of $(1 - x)$, then x equals

(A) -2　　(B) -1　　(C) $1/2$　　(D) 2　　(E) 3

2. For how many real numbers x is $\sqrt{-(x + 1)^2}$ a real number?

(A) none　　(B) one　　(C) two
(D) a finite number greater than two　　(E) infinitely many

3. The sum of the distances from one vertex of a square with sides of length two to the midpoints of each of the sides of the square is

(A) $2\sqrt{5}$　　(B) $2 + \sqrt{3}$　　(C) $2 + 2\sqrt{3}$　　(D) $2 + \sqrt{5}$
(E) $2 + 2\sqrt{5}$

4. Let a geometric progression with n terms have first term one, common ratio r and sum s, where r and s are not zero. The sum of the geometric progression formed by replacing each term of the original progression by its reciprocal is

(A) $\dfrac{1}{s}$　　(B) $\dfrac{1}{r^n s}$　　(C) $\dfrac{s}{r^{n-1}}$　　(D) $\dfrac{r^n}{s}$　　(E) $\dfrac{r^{n-1}}{s}$

5. How many integers greater than ten and less than one hundred, written in base ten notation, are increased by nine when their digits are reversed?

(A) 0　　(B) 1　　(C) 8　　(D) 9　　(E) 10

6. If c is a real number and the negative of one of the solutions of $x^2 - 3x + c = 0$ is a solution of $x^2 + 3x - c = 0$, then the solutions of $x^2 - 3x + c = 0$ are

(A) $1, 2$　　(B) $-1, -2$　　(C) $0, 3$　　(D) $0, -3$　　(E) $\dfrac{3}{2}, \dfrac{3}{2}$

7. If x is a real number, then the quantity $(1 - |x|)(1 + x)$ is positive if and only if

(A) $|x| < 1$　　(B) $x < 1$　　(C) $|x| > 1$　　(D) $x < -1$
(E) $x < -1$ or $-1 < x < 1$

8. A point in the plane, both of whose rectangular coordinates are integers with absolute value less than or equal to four, is chosen at random, with all such points having an equal probability of being chosen. What is the probability that the distance from the point to the origin is at most two units?

(A) $\dfrac{13}{81}$ (B) $\dfrac{15}{81}$ (C) $\dfrac{13}{64}$ (D) $\dfrac{\pi}{16}$

(E) the square of a rational number

9. In triangle ABC, D is the midpoint of AB; E is the midpoint of DB; and F is the midpoint of BC. If the area of $\triangle ABC$ is 96, then the area of $\triangle AEF$ is

(A) 16   (B) 24   (C) 32   (D) 36   (E) 48

10. If m, n, p and q are real numbers and $f(x) = mx + n$ and $g(x) = px + q$, then the equation $f(g(x)) = g(f(x))$ has a solution

(A) for all choices of m, n, p and q
(B) if and only if $m = p$ and $n = q$
(C) if and only if $mq - np = 0$
(D) if and only if $n(1 - p) - q(1 - m) = 0$
(E) if and only if $(1 - n)(1 - p) - (1 - q)(1 - m) = 0$

11. Which of the following statements is (are) equivalent to the statement "If the pink elephant on planet alpha has purple eyes, then the wild pig on planet beta does not have a long nose"?

    I. "If the wild pig on planet beta has a long nose, then the pink elephant on planet alpha has purple eyes."
   II. "If the pink elephant on planet alpha does not have purple eyes, then the wild pig on planet beta does not have a long nose."
  III. "If the wild pig on planet beta has a long nose, then the pink elephant on planet alpha does not have purple eyes."
  IV. "The pink elephant on planet alpha does not have purple eyes, or[†] the wild pig on planet beta does not have a long nose."

(A) I and III only   (B) III and IV only
(C) II and IV only   (D) II and III only   (E) III only

[†]The word "or" is used here in the inclusive sense (as is customary in mathematical writing).

12. A supermarket has 128 crates of apples. Each crate contains at least 120 apples and at most 144 apples. What is the largest integer n such that there must be at least n crates containing the same number of apples?

(A) 4   (B) 5   (C) 6   (D) 24   (E) 25

13. If x cows give $x + 1$ cans of milk in $x + 2$ days, how many days will it take $x + 3$ cows to give $x + 5$ cans of milk?

(A) $\dfrac{x(x + 2)(x + 5)}{(x + 1)(x + 3)}$ (B) $\dfrac{x(x + 1)(x + 5)}{(x + 2)(x + 3)}$

(C) $\dfrac{(x + 1)(x + 3)(x + 5)}{x(x + 2)}$ (D) $\dfrac{(x + 1)(x + 3)}{x(x + 2)(x + 5)}$

(E) none of these

14. The measures of the interior angles of a convex polygon are in arithmetic progression. If the smallest angle is $100°$ and the largest angle is $140°$, then the number of sides the polygon has is

(A) 6   (B) 8   (C) 10   (D) 11   (E) 12

15. If r is the remainder when each of the numbers 1059, 1417 and 2312 is divided by d, where d is an integer greater than one, then $d - r$ equals

(A) 1   (B) 15   (C) 179   (D) $d - 15$ (E) $d - 1$

16. In triangles ABC and DEF, lengths AC, BC, DF and EF are all equal. Length AB is twice the length of the altitude of $\triangle DEF$ from F to DE. Which of the following statements is (are) true?

I. $\angle ACB$ and $\angle DFE$ must be complementary.
II. $\angle ACB$ and $\angle DFE$ must be supplementary.
III. The area of $\triangle ABC$ must equal the area of $\triangle DEF$.
IV. The area of $\triangle ABC$ must equal twice the area of $\triangle DEF$.

(A) II only   (B) III only   (C) IV only
(D) I and III only   (E) II and III only

17. If θ is an acute angle and $\sin 2\theta = a$, then $\sin \theta + \cos \theta$ equals

(A) $\sqrt{a + 1}$ (B) $(\sqrt{2} - 1)a + 1$ (C) $\sqrt{a + 1} - \sqrt{a^2 - a}$

(D) $\sqrt{a + 1} + \sqrt{a^2 - a}$ (E) $\sqrt{a + 1} + a^2 - a$

18. In the adjoining figure, AB is tangent at A to the circle with center O; point D is interior to the circle; and DB intersects the circle at C. If $BC = DC = 3$, $OD = 2$ and $AB = 6$, then the radius of the circle is

(A) $3 + \sqrt{3}$

(B) $15/\pi$

(C) $9/2$

(D) $2\sqrt{6}$

(E) $\sqrt{22}$

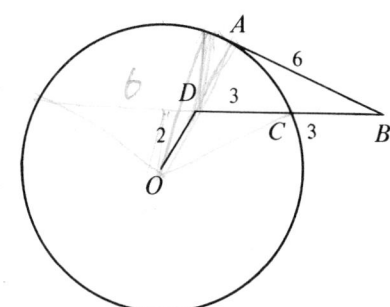

19. A polynomial $p(x)$ has remainder three when divided by $x - 1$ and remainder five when divided by $x - 3$. The remainder when $p(x)$ is divided by $(x - 1)(x - 3)$ is

(A) $x - 2$ (B) $x + 2$ (C) 2   (D) 8   (E) 15

20. Let a, b and x be positive real numbers distinct from one. Then

$$4(\log_a x)^2 + 3(\log_b x)^2 = 8(\log_a x)(\log_b x)$$

(A) for all values of a, b and x (B) if and only if $a = b^2$
(C) if and only if $b = a^2$ (D) if and only if $x = ab$
(E) none of these

21. What is the smallest positive odd integer n such that the product

$$2^{1/7} 2^{3/7} \cdots 2^{(2n+1)/7}$$

is greater than 1000? (In the product the denominators of the exponents are all sevens, and the numerators are the successive odd integers from 1 to $2n + 1$.)

(A) 7   (B) 9   (C) 11   (D) 17   (E) 19

22. Given an equilateral triangle with side of length s, consider the locus of all points P in the plane of the triangle such that the sum of the squares of the distances from P to the vertices of the triangle is a fixed number a. This locus

   (A) is a circle if $a > s^2$
   (B) contains only three points if $a = 2s^2$ and is a circle if $a > 2s^2$
   (C) is a circle with positive radius only if $s^2 < a < 2s^2$
   (D) contains only a finite number of points for any value of a
   (E) is none of these

23. For integers k and n such that $1 \leqslant k < n$, let
$$\binom{n}{k} = \frac{n!}{k!(n-k)!}.$$
Then $\left(\dfrac{n-2k-1}{k+1}\right)\dbinom{n}{k}$ is an integer

   (A) for all k and n
   (B) for all even values of k and n, but not for all k and n
   (C) for all odd values of k and n, but not for all k and n
   (D) if $k = 1$ or $n - 1$, but not for all odd values of k and n
   (E) if n is divisible by k, but not for all even values of k and n

24. In the adjoining figure, circle K has diameter AB; circle L is tangent to circle K and to AB at the center of circle K; and circle M is tangent to circle K, to circle L and to AB. The ratio of the area of circle K to the area of circle M is

   (A) 12

   (B) 14

   (C) 16

   (D) 18

   (E) not an integer

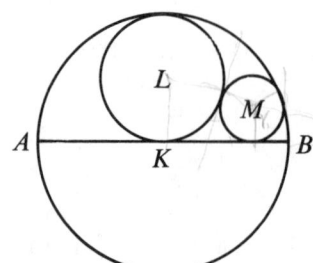

25. For a sequence $u_1, u_2, \ldots$, define $\Delta^1(u_n) = u_{n+1} - u_n$ and, for all integers $k > 1, \Delta^k(u_n) = \Delta^1(\Delta^{k-1}(u_n))$. If $u_n = n^3 + n$, then $\Delta^k(u_n) = 0$ for all n

   (A) if $k = 1$ (B) if $k = 2$, but not if $k = 1$
   (C) if $k = 3$, but not if $k = 2$ (D) if $k = 4$, but not if $k = 3$
   (E) for no value of k

26. In the adjoining figure, every point of circle O' is exterior to circle O. Let P and Q be the points of intersection of an internal common tangent with the two external common tangents. Then the length of PQ is

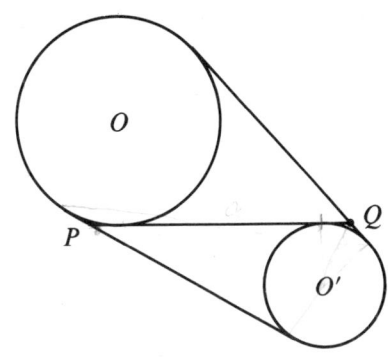

(A) the average of the lengths of the internal and external common tangents

(B) equal to the length of an external common tangent if and only if circles O and O' have equal radii

(C) always equal to the length of an external common tangent

(D) greater than the length of an external common tangent

(E) the geometric mean of the lengths of the internal and external common tangents

27. If

$$N = \frac{\sqrt{\sqrt{5} + 2} + \sqrt{\sqrt{5} - 2}}{\sqrt{\sqrt{5} + 1}} - \sqrt{3 - 2\sqrt{2}},$$

then N equals

(A) 1   (B) $2\sqrt{2} - 1$ (C) $\dfrac{\sqrt{5}}{2}$ (D) $\sqrt{\dfrac{5}{2}}$

(E) none of these

28. Lines $L_1, L_2 \ldots, L_{100}$ are distinct. All lines L_{4n}, n a positive integer, are parallel to each other. All lines L_{4n-3}, n a positive integer, pass through a given point A. The maximum number of points of intersection of pairs of lines from the complete set $\{L_1, L_2, \ldots, L_{100}\}$ is

(A) 4350   (B) 4351   (C) 4900   (D) 4901   (E) 9851

29. Ann and Barbara were comparing their ages and found that Barbara is as old as Ann was when Barbara was as old as Ann had been when Barbara was half as old as Ann is. If the sum of their present ages is 44 years, then Ann's age is

   (A) 22   (B) 24   (C) 25   (D) 26   (E) 28

30. How many distinct ordered triples (x, y, z) satisfy the equations

$$x + 2y + 4z = 12$$
$$xy + 4yz + 2xz = 22$$
$$xyz = 6 \ ?$$

   (A) none   (B) 1   (C) 2   (D) 4   (E) 6

1977 Examination

1. If $y = 2x$ and $z = 2y$, then $x + y + z$ equals

   (A) x (B) $3x$ (C) $5x$ (D) $7x$ (E) $9x$

2. Which one of the following statements is false? All equilateral triangles are

   (A) equiangular   (B) isosceles   (C) regular polygons
   (D) congruent to each other   (E) similar to each other

3. A man has \$2.73 in pennies, nickels, dimes, quarters and half dollars. If he has an equal number of coins of each kind, then the total number of coins he has is

   (A) 3   (B) 5   (C) 9   (D) 10   (E) 15

4. In triangle ABC, $AB = AC$ and $\angle A = 80°$. If points D, E and F lie on sides BC, AC and AB, respectively, and $CE = CD$ and $BF = BD$, then $\angle EDF$ equals

   (A) 30°   (B) 40°
   (C) 50°   (D) 65°
   (E) none of these

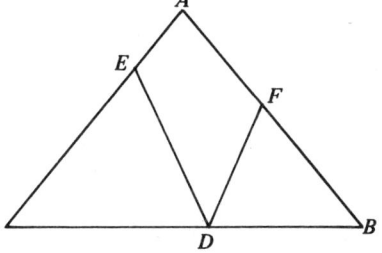

5. The set of all points P such that the sum of the (undirected) distances from P to two fixed points A and B equals the distance between A and B is

   (A) the line segment from A to B
   (B) the line passing through A and B
   (C) the perpendicular bisector of the line segment from A to B
   (D) an ellipse having positive area
   (E) a parabola

6. If x, y and $2x + \dfrac{y}{2}$ are not zero, then

$$\left(2x + \frac{y}{2}\right)^{-1}\left[(2x)^{-1} + \left(\frac{y}{2}\right)^{-1}\right]$$

equals

(A) 1   (B) xy^{-1} (C) $x^{-1}y$ (D) $(xy)^{-1}$
(E) none of these

7. If $t = \dfrac{1}{1 - \sqrt[4]{2}}$, then t equals

(A) $(1 - \sqrt[4]{2})(2 - \sqrt{2})$ (B) $(1 - \sqrt[4]{2})(1 + \sqrt{2})$
(C) $(1 + \sqrt[4]{2})(1 - \sqrt{2})$ (D) $(1 + \sqrt[4]{2})(1 + \sqrt{2})$
(E) $-(1 + \sqrt[4]{2})(1 + \sqrt{2})$

8. For every triple (a, b, c) of non-zero real numbers, form the number

$$\frac{a}{|a|} + \frac{b}{|b|} + \frac{c}{|c|} + \frac{abc}{|abc|}.$$

The set of all numbers formed is

(A) $\{0\}$ (B) $\{-4, 0, 4\}$ (C) $\{-4, -2, 0, 2, 4\}$
(D) $\{-4, -2, 2, 4\}$ (E) none of these

9. In the adjoining figure $\measuredangle E = 40°$ and arc AB, arc BC and arc CD all have equal length. Find the measure of $\measuredangle ACD$.

(A) $10°$ (B) $15°$
(C) $20°$ (D) $\left(\dfrac{45}{2}\right)°$
(E) $30°$

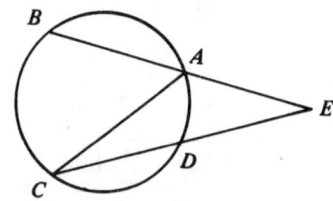

10. If $(3x - 1)^7 = a_7x^7 + a_6x^6 + \cdots + a_0$, then $a_7 + a_6 + \cdots + a_0$ equals

(A) 0   (B) 1   (C) 64   (D) -64 (E) 128

11. For each real number x, let $[x]$ be the largest integer not exceeding x (i.e., the integer n such that $n \leqslant x < n + 1$). Which of the following statements is (are) true?

   I. $[x + 1] = [x] + 1$ for all x
   II. $[x + y] = [x] + [y]$ for all x and y
   III. $[xy] = [x][y]$ for all x and y

   (A) none   (B) I only   (C) I and II only
   (D) III only   (E) all

12. Al's age is 16 more than the sum of Bob's age and Carl's age, and the square of Al's age is 1632 more than the square of the sum of Bob's age and Carl's age. The sum of the ages of Al, Bob and Carl is

   (A) 64   (B) 94   (C) 96   (D) 102   (E) 140

13. If $a_1, a_2, a_3, \ldots$ is a sequence of positive numbers such that $a_{n+2} = a_n a_{n+1}$ for all positive integers n, then the sequence $a_1, a_2, a_3, \ldots$ is a geometric progression

   (A) for all positive values of a_1 and a_2
   (B) if and only if $a_1 = a_2$ (C) if and only if $a_1 = 1$
   (D) if and only if $a_2 = 1$ (E) if and only if $a_1 = a_2 = 1$

14. How many pairs (m, n) of integers satisfy the equation $m + n = mn$?

   (A) 1   (B) 2   (C) 3   (D) 4   (E) more than 4

15. Each of the three circles in the adjoining figure is externally tangent to the other two, and each side of the triangle is tangent to two of the circles. If each circle has radius three, then the perimeter of the triangle is

   (A) $36 + 9\sqrt{2}$ (B) $36 + 6\sqrt{3}$

   (C) $36 + 9\sqrt{3}$ (D) $18 + 18\sqrt{3}$

   (E) 45

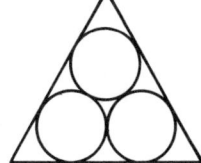

16. If $i^2 = -1$, then the sum

$$\cos 45° + i\cos 135° + \cdots + i^n\cos(45 + 90n)° + \cdots + i^{40}\cos 3645°$$

equals

(A) $\dfrac{\sqrt{2}}{2}$ (B) $-10i\sqrt{2}$ (C) $\dfrac{21\sqrt{2}}{2}$

(D) $\dfrac{\sqrt{2}}{2}(21 - 20i)$ (E) $\dfrac{\sqrt{2}}{2}(21 + 20i)$

17. Three fair dice are tossed (all faces have the same probability of coming up). What is the probability that the three numbers turned up can be arranged to form an arithmetic progression with common difference one?

(A) $\dfrac{1}{6}$ (B) $\dfrac{1}{9}$ (C) $\dfrac{1}{27}$ (D) $\dfrac{1}{54}$ (E) $\dfrac{7}{36}$

18. If $y = (\log_2 3)(\log_3 4) \cdots (\log_n[n + 1]) \cdots (\log_{31} 32)$ then

(A) $4 < y < 5$ (B) $y = 5$ (C) $5 < y < 6$
(D) $y = 6$ (E) $6 < y < 7$

19. Let E be the point of intersection of the diagonals of convex quadrilateral $ABCD$, and let P, Q, R and S be the centers of the circles circumscribing triangles ABE, BCE, CDE and ADE, respectively. Then

(A) $PQRS$ is a parallelogram
(B) $PQRS$ is a parallelogram if and only if $ABCD$ is a rhombus
(C) $PQRS$ is a parallelogram if and only if $ABCD$ is a rectangle
(D) $PQRS$ is a parallelogram if and only if $ABCD$ is a parallelogram
(E) none of the above are true

20. For how many paths consisting of a sequence of horizontal and/or vertical line segments, with each segment connecting a pair of adjacent letters in the diagram below, is the word CONTEST spelled out as the path is traversed from beginning to end?

(A) 63   (B) 128
(C) 129   (D) 255
(E) none of these

C
COC
CONOC
CONTNOC
CONTETNOC
CONTESETNOC
CONTESTSETNOC

21. For how many values of the coefficient a do the equations

$$x^2 + ax + 1 = 0,$$
$$x^2 - x - a = 0$$

have a common real solution?

(A) 0   (B) 1   (C) 2   (D) 3   (E) infinitely many

22. If $f(x)$ is a real valued function of the real variable x, and $f(x)$ is not identically zero, and for all a and b

$$f(a + b) + f(a - b) = 2f(a) + 2f(b),$$

then for all x and y

(A) $f(0) = 1$ (B) $f(-x) = -f(x)$ (C) $f(-x) = f(x)$
(D) $f(x + y) = f(x) + f(y)$ (E) there is a positive number T such that $f(x + T) = f(x)$

23. If the solutions of the equation $x^2 + px + q = 0$ are the cubes of the solutions of the equation $x^2 + mx + n = 0$, then

(A) $p = m^3 + 3mn$ (B) $p = m^3 - 3mn$ (C) $p + q = m^3$
(D) $\left(\dfrac{m}{n}\right)^3 = \dfrac{p}{q}$ (E) none of these

24. Find the sum

$$\frac{1}{1 \cdot 3} + \frac{1}{3 \cdot 5} + \cdots + \frac{1}{(2n - 1)(2n + 1)} + \cdots + \frac{1}{255 \cdot 257}.$$

(A) $\dfrac{127}{255}$ (B) $\dfrac{128}{255}$ (C) $\dfrac{1}{2}$ (D) $\dfrac{128}{257}$ (E) $\dfrac{129}{257}$

25. Determine the largest positive integer n such that $1005!$ is divisible by 10^n.

(A) 102   (B) 112   (C) 249   (D) 502   (E) none of these

26. Let a, b, c and d be the lengths of sides MN, NP, PQ and QM, respectively, of quadrilateral $MNPQ$. If A is the area of $MNPQ$, then

(A) $A = \left(\dfrac{a + c}{2}\right)\left(\dfrac{b + d}{2}\right)$ if and only if $MNPQ$ is convex

(B) $A = \left(\dfrac{a + c}{2}\right)\left(\dfrac{b + d}{2}\right)$ if and only if $MNPQ$ is a rectangle

(C) $A \leqslant \left(\dfrac{a + c}{2}\right)\left(\dfrac{b + d}{2}\right)$ if and only if $MNPQ$ is a rectangle

(D) $A \leqslant \left(\dfrac{a + c}{2}\right)\left(\dfrac{b + d}{2}\right)$ if and only if $MNPQ$ is a parallelogram

(E) $A \geqslant \left(\dfrac{a + c}{2}\right)\left(\dfrac{b + d}{2}\right)$ if and only if $MNPQ$ is a parallelogram

27. There are two spherical balls of different sizes lying in two corners of a rectangular room, each touching two walls and the floor. If there is a point on each ball which is 5 inches from each wall which that ball touches and 10 inches from the floor, then the sum of the diameters of the balls is

(A) 20 inches   (B) 30 inches   (C) 40 inches
(D) 60 inches   (E) not determined by the given information

28. Let $g(x) = x^5 + x^4 + x^3 + x^2 + x + 1$. What is the remainder when the polynomial $g(x^{12})$ is divided by the polynomial $g(x)$?

(A) 6   (B) $5 - x$ (C) $4 - x + x^2$ (D) $3 - x + x^2 - x^3$
(E) $2 - x + x^2 - x^3 + x^4$

29. Find the smallest integer n such that

$$\left(x^2 + y^2 + z^2\right)^2 \leqslant n\left(x^4 + y^4 + z^4\right)$$

for all real numbers $x, y,$ and z.

(A) 2 (B) 3 (C) 4 (D) 6 (E) There is no such integer n.

30. If a, b and d are the lengths of a side, a shortest diagonal and a longest diagonal, respectively, of a regular nonagon (see adjoining figure), then

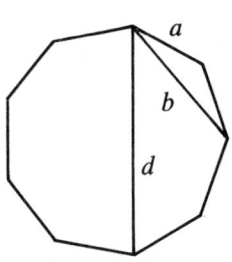

(A) $d = a + b$

(B) $d^2 = a^2 + b^2$

(C) $d^2 = a^2 + ab + b^2$

(D) $b = \dfrac{a + d}{2}$

(E) $b^2 = ad$

1978 Examination

1. If $1 - \dfrac{4}{x} + \dfrac{4}{x^2} = 0,$ then $\dfrac{2}{x}$ equals

   (A) -1 (B) 1 (C) 2 (D) -1 or 2 (E) -1 or -2

2. If four times the reciprocal of the circumference of a circle equals the diameter of the circle, then the area of the circle is

   (A) $\dfrac{1}{\pi^2}$ (B) $\dfrac{1}{\pi}$ (C) 1 (D) π (E) π^2

3. For all non-zero numbers x and y such that $x = 1/y$,

   $$\left(x - \frac{1}{x}\right)\left(y + \frac{1}{y}\right)$$

   equals

   (A) $2x^2$ (B) $2y^2$ (C) $x^2 + y^2$ (D) $x^2 - y^2$ (E) $y^2 - x^2$

4. If $a = 1, b = 10, c = 100$ and $d = 1000,$ then

   $$(a + b + c - d) + (a + b - c + d)$$
   $$+ (a - b + c + d) + (-a + b + c + d)$$

   is equal to

   (A) 1111 (B) 2222 (C) 3333 (D) 1212 (E) 4242

5. Four boys bought a boat for \$60. The first boy paid one half of the sum of the amounts paid by the other boys; the second boy paid one third of the sum of the amounts paid by the other boys; and the third boy paid one fourth of the sum of the amounts paid by the other boys. How much did the fourth boy pay?

   (A) \$10   (B) \$12   (C) \$13   (D) \$14   (E) \$15

6. The number of distinct pairs (x, y) of real numbers satisfying both of the following equations:

$$x = x^2 + y^2,$$
$$y = 2xy$$

is

(A) 0   (B) 1   (C) 2   (D) 3   (E) 4

7. Opposite sides of a regular hexagon are 12 inches apart. The length of each side, in inches, is

(A) 7.5   (B) $6\sqrt{2}$ (C) $5\sqrt{2}$ (D) $\frac{9}{2}\sqrt{3}$ (E) $4\sqrt{3}$

8. If $x \neq y$ and the sequences x, a_1, a_2, y and x, b_1, b_2, b_3, y each are in arithmetic progression, then $(a_2 - a_1)/(b_2 - b_1)$ equals

(A) $\frac{2}{3}$ (B) $\frac{3}{4}$ (C) 1   (D) $\frac{4}{3}$ (E) $\frac{3}{2}$

9. If $x < 0$, then $|x - \sqrt{(x - 1)^2}|$ equals

(A) 1   (B) $1 - 2x$ (C) $-2x - 1$ (D) $1 + 2x$ (E) $2x - 1$

10. If B is a point on circle C with center P, then the set of all points A in the plane of circle C such that the distance between A and B is less than or equal to the distance between A and any other point on circle C is

(A) the line segment from P to B
(B) the ray beginning at P and passing through B
(C) a ray beginning at B
(D) a circle whose center is P
(E) a circle whose center is B

11. If r is positive and the line whose equation is $x + y = r$ is tangent to the circle whose equation is $x^2 + y^2 = r$, then r equals

(A) $\frac{1}{2}$ (B) 1   (C) 2   (D) $\sqrt{2}$ (E) $2\sqrt{2}$

12. In $\triangle ADE, \angle ADE = 140°$ points B and C lie on sides AD and AE, respectively, and points A, B, C, D, E are distinct.[†] If lengths AB, BC, CD and DE are all equal, then the measure of $\angle EAD$ is

(A) 5°　　(B) 6°　　(C) 7.5°　　(D) 8°　　(E) 10°

13. If a, b, c, and d are non-zero numbers such that c and d are the solutions of $x^2 + ax + b = 0$ and a and b are the solutions of $x^2 + cx + d = 0$, then $a + b + c + d$ equals

(A) 0　　(B) -2　　(C) 2　　(D) 4　　(E) $(-1 + \sqrt{5})/2$

14. If an integer n, greater than 8, is a solution of the equation $x^2 - ax + b = 0$ and the representation of a in the base n numeration system is 18, then the base n representation of b is

(A) 18　　(B) 28　　(C) 80　　(D) 81　　(E) 280

15. If $\sin x + \cos x = 1/5$ and $0 \leqslant x < \pi$, then $\tan x$ is

(A) $-\dfrac{4}{3}$　　(B) $-\dfrac{3}{4}$　　(C) $\dfrac{3}{4}$　　(D) $\dfrac{4}{3}$

(E) not completely determined by the given information

16. In a room containing N people, $N > 3$, at least one person has not shaken hands with everyone else in the room. What is the maximum number of people in the room that could have shaken hands with everyone else?

(A) 0　　(B) 1　　(C) $N - 1$　　(D) N　　(E) none of these

17. If k is a positive number and f is a function such that, for every positive number x,

$$\left[f(x^2 + 1)\right]^{\sqrt{x}} = k;$$

then, for every positive number y,

$$\left[f\left(\frac{9 + y^2}{y^2}\right)\right]^{\sqrt{\frac{12}{y}}}$$

is equal to

(A) $\sqrt{k}$　　(B) $2k$　　(C) $k\sqrt{k}$　　(D) k^2　　(E) $y\sqrt{k}$

[†] The specification that points A, B, C, D, E be distinct was not included in the original statement of the problem. If $B = D$, then $C = E$ and $\angle EAD = 20°$.

18. What is the smallest positive integer n such that $\sqrt{n} - \sqrt{n-1}$ < .01?

(A) 2499   (B) 2500   (C) 2501   (D) 10,000
(E) There is no such integer.

19. A positive integer n not exceeding 100 is chosen in such a way that if $n \leqslant 50$, then the probability of choosing n is p, and if $n > 50$, then the probability of choosing n is $3p$. The probability that a perfect square is chosen is

(A) .05   (B) .065   (C) .08   (D) .09   (E) .1

20. If a, b, c are non-zero real numbers such that

$$\frac{a+b-c}{c} = \frac{a-b+c}{b} = \frac{-a+b+c}{a},$$

and

$$x = \frac{(a+b)(b+c)(c+a)}{abc},$$

and $x < 0$, then x equals

(A) -1 (B) -2 (C) -4 (D) -6 (E) -8

21. For all positive numbers x distinct from 1,

$$\frac{1}{\log_3 x} + \frac{1}{\log_4 x} + \frac{1}{\log_5 x}$$

equals

(A) $\dfrac{1}{\log_{60} x}$ (B) $\dfrac{1}{\log_x 60}$ (C) $\dfrac{1}{(\log_3 x)(\log_4 x)(\log_5 x)}$

(D) $\dfrac{12}{(\log_3 x) + (\log_4 x) + (\log_5 x)}$

(E) $\dfrac{\log_2 x}{(\log_3 x)(\log_5 x)} + \dfrac{\log_3 x}{(\log_2 x)(\log_5 x)} + \dfrac{\log_5 x}{(\log_2 x)(\log_3 x)}$

22. The following four statements, and only these, are found on a card:

> On this card exactly one statement is false.
> On this card exactly two statements are false.
> On this card exactly three statements are false.
> On this card exactly four statements are false.

(Assume each statement on the card is either true or false.) Among them the number of false statements is exactly

(A) 0   (B) 1   (C) 2   (D) 3   (E) 4

23. Vertex E of equilateral triangle ABE is in the interior of square $ABCD$, and F is the point of intersection of diagonal BD and line segment AE. If length AB is $\sqrt{1 + \sqrt{3}}$ then the area of $\triangle ABF$ is

(A) 1   (B) $\dfrac{\sqrt{2}}{2}$ (C) $\dfrac{\sqrt{3}}{2}$

(D) $4 - 2\sqrt{3}$ (E) $\dfrac{1}{2} + \dfrac{\sqrt{3}}{4}$

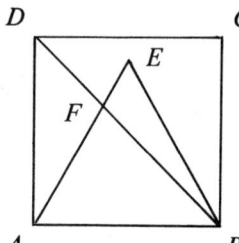

24. If the distinct non-zero numbers $x(y - z)$, $y(z - x)$, $z(x - y)$ form a geometric progression with common ratio r, then r satisfies the equation

(A) $r^2 + r + 1 = 0$ (B) $r^2 - r + 1 = 0$ (C) $r^4 + r^2 - 1 = 0$
(D) $(r + 1)^4 + r = 0$ (E) $(r - 1)^4 + r = 0$

25. Let a be a positive number. Consider the set S of all points whose rectangular coordinates (x, y) satisfy all of the following conditions:

(i) $\dfrac{a}{2} \leqslant x \leqslant 2a$ (ii) $\dfrac{a}{2} \leqslant y \leqslant 2a$ (iii) $x + y \geqslant a$

(iv) $x + a \geqslant y$ (v) $y + a \geqslant x$

The boundary of set S is a polygon with

(A) 3 sides   (B) 4 sides   (C) 5 sides   (D) 6 sides
(E) 7 sides

26. In $\triangle ABC$, $AB = 10$, $AC = 8$ and $BC = 6$. Circle P is the circle with smallest radius which passes through C and is tangent to AB. Let Q and R be the points of intersection, distinct from C, of circle P with sides AC and BC, respectively. The length of segment QR is

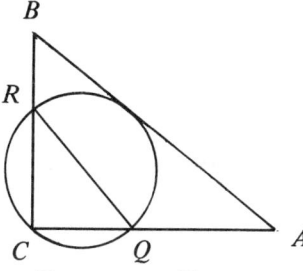

(A) 4.75   (B) 4.8   (C) 5   (D) $4\sqrt{2}$ (E) $3\sqrt{3}$

27. There is more than one integer greater than 1 which, when divided by any integer k such that $2 \leqslant k \leqslant 11$, has a remainder of 1. What is the difference between the two smallest such integers?

(A) 2310   (B) 2311   (C) 27,720   (D) 27,721
(E) none of these

28. If $\triangle A_1 A_2 A_3$ is equilateral and A_{n+3} is the midpoint of line segment $A_n A_{n+1}$ for all positive integers n, then the measure of $\measuredangle A_{44} A_{45} A_{43}$ equals

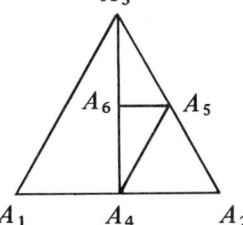

(A) 30°   (B) 45°   (C) 60°
(D) 90°   (E) 120°

29. Sides AB, BC, CD and DA, respectively, of convex quadrilateral $ABCD$ are extended past B, C, D and A to points B', C', D' and A'. Also, $AB = BB' = 6$, $BC = CC' = 7$, $CD = DD' = 8$ and $DA = AA' = 9$; and the area of $ABCD$ is 10. The area of $A'B'C'D'$ is

(A) 20   (B) 40   (C) 45   (D) 50   (E) 60

30. In a tennis tournament, n women and $2n$ men play, and each player plays exactly one match with every other player. If there are no ties and the ratio of the number of matches won by women to the number of matches won by men is $7/5$, then n equals

(A) 2   (B) 4   (C) 6   (D) 7   (E) none of these

1979 Examination

1. If rectangle $ABCD$ has area 72 square meters and E and G are the midpoints of sides AD and CD, respectively, then the area of rectangle $DEFG$ in square meters is

   (A) 8   (B) 9   (C) 12

   (D) 18   (E) 24

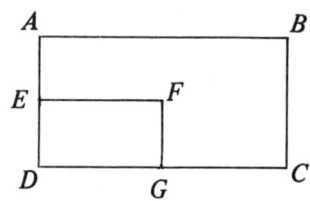

2. For all non-zero real numbers x and y such that $x - y = xy$, $\dfrac{1}{x} - \dfrac{1}{y}$ equals

   (A) $\dfrac{1}{xy}$ (B) $\dfrac{1}{x - y}$ (C) 0   (D) -1 (E) $y - x$

3. In the adjoining figure, $ABCD$ is a square, ABE is an equilateral triangle and point E is outside square $ABCD$. What is the measure of $\angle AED$ in degrees?

   (A) 10

   (B) 12.5

   (C) 15

   (D) 20

   (E) 25

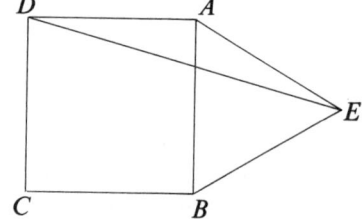

4. For all real numbers x, $x[x\{x(2 - x) - 4\} + 10] + 1 =$

   (A) $-x^4 + 2x^3 + 4x^2 + 10x + 1$
   (B) $-x^4 - 2x^3 + 4x^2 + 10x + 1$
   (C) $-x^4 - 2x^3 - 4x^2 + 10x + 1$
   (D) $-x^4 - 2x^3 - 4x^2 - 10x + 1$
   (E) $-x^4 + 2x^3 - 4x^2 + 10x + 1$

5. Find the sum of the digits of the largest even three digit number (in base ten representation) which is not changed when its units and hundreds digits are interchanged.

   (A) 22   (B) 23   (C) 24   (D) 25   (E) 26

6. $\dfrac{3}{2} + \dfrac{5}{4} + \dfrac{9}{8} + \dfrac{17}{16} + \dfrac{33}{32} + \dfrac{65}{64} - 7 =$

(A) $-\dfrac{1}{64}$ (B) $-\dfrac{1}{16}$ (C) 0 (D) $\dfrac{1}{16}$ (E) $\dfrac{1}{64}$

7. The square of an integer is called a *perfect square*. If x is a perfect square, the next larger perfect square is

(A) $x + 1$ (B) $x^2 + 1$ (C) $x^2 + 2x + 1$ (D) $x^2 + x$

(E) $x + 2\sqrt{x} + 1$

8. Find the area of the smallest region bounded by the graphs of $y = |x|$ and $x^2 + y^2 = 4$.

(A) $\dfrac{\pi}{4}$ (B) $\dfrac{3\pi}{4}$ (C) π (D) $\dfrac{3\pi}{2}$ (E) 2π

9. The product of $\sqrt[3]{4}$ and $\sqrt[4]{8}$ equals

(A) $\sqrt[7]{12}$ (B) $2\sqrt[7]{12}$ (C) $\sqrt[7]{32}$ (D) $\sqrt[12]{32}$ (E) $2\sqrt[12]{32}$

10. If $P_1P_2P_3P_4P_5P_6$ is a regular hexagon whose apothem (distance from the center to the midpoint of a side) is 2, and Q_i is the midpoint of side P_iP_{i+1} for $i = 1, 2, 3, 4$, then the area of quadrilateral $Q_1Q_2Q_3Q_4$ is

(A) 6 (B) $2\sqrt{6}$ (C) $\dfrac{8\sqrt{3}}{3}$ (D) $3\sqrt{3}$ (E) $4\sqrt{3}$

11. Find a positive integral solution to the equation

$$\frac{1 + 3 + 5 + \cdots + (2n - 1)}{2 + 4 + 6 + \cdots + 2n} = \frac{115}{116}.$$

(A) 110 (B) 115 (C) 116 (D) 231
(E) The equation has no positive integral solutions.

12. In the adjoining figure, CD is the diameter of a semi-circle with center O. Point A lies on the extension of DC past C; point E lies on the semi-circle, and B is the point of intersection (distinct from E) of line segment AE with the semi-circle. If length AB equals length OD, and the measure of $\angle EOD$ is $45°$, then the measure of $\angle BAO$ is

(A) $10°$ (B) $15°$

(C) $20°$ (D) $25°$

(E) $30°$

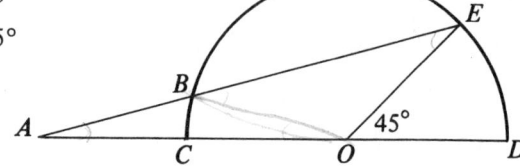

13. The inequality $y - x < \sqrt{x^2}$ is satisfied if and only if

(A) $y < 0$ or $y < 2x$ (or both inequalities hold)
(B) $y > 0$ or $y < 2x$ (or both inequalities hold)
(C) $y^2 < 2xy$ (D) $y < 0$ (E) $x > 0$ and $y < 2x$

14. In a certain sequence of numbers, the first number is 1, and, for all $n \geqslant 2$, the product of the first n numbers in the sequence is n^2. The sum of the third and the fifth numbers in the sequence is

(A) $\dfrac{25}{9}$ (B) $\dfrac{31}{15}$ (C) $\dfrac{61}{16}$ (D) $\dfrac{576}{225}$ (E) 34

15. Two identical jars are filled with alcohol solutions, the ratio of the volume of alcohol to the volume of water being $p : 1$ in one jar and $q : 1$ in the other jar. If the entire contents of the two jars are mixed together, the ratio of the volume of alcohol to the volume of water in the mixture is

(A) $\dfrac{p + q}{2}$ (B) $\dfrac{p^2 + q^2}{p + q}$ (C) $\dfrac{2pq}{p + q}$

(D) $\dfrac{2(p^2 + pq + q^2)}{3(p + q)}$ (E) $\dfrac{p + q + 2pq}{p + q + 2}$

16. A circle with area A_1 is contained in the interior of a larger circle with area $A_1 + A_2$. If the radius of the larger circle is 3, and if $A_1, A_2, A_1 + A_2$ is an arithmetic progression, then the radius of the smaller circle is

(A) $\dfrac{\sqrt{3}}{2}$ (B) 1 (C) $\dfrac{2}{\sqrt{3}}$ (D) $\dfrac{3}{2}$ (E) $\sqrt{3}$

17. Points A, B, C and D are distinct and lie, in the given order, on a straight line. Line segments AB, AC and AD have lengths x, y and z, respectively. If line segments AB and CD may be rotated about points B and C, respectively, so that points A and D coincide, to form a triangle with positive area, then which of the following three inequalities must be satisfied?

I. $x < \dfrac{z}{2}$

II. $y < x + \dfrac{z}{2}$

III. $y < \dfrac{z}{2}$

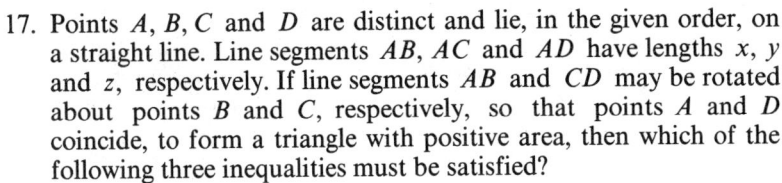
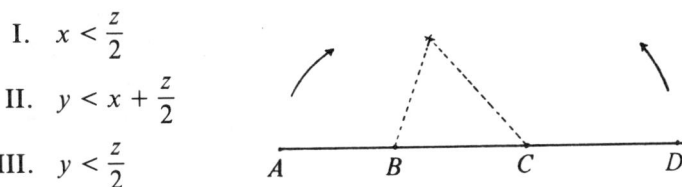

(A) I only   (B) II only   (C) I and II only
(D) II and III only   (E) I, II and III

18. To the nearest thousandth, $\log_{10}2$ is .301 and $\log_{10}3$ is .477. Which of the following is the best approximation of $\log_5 10$?

(A) $\dfrac{8}{7}$ (B) $\dfrac{9}{7}$ (C) $\dfrac{10}{7}$ (D) $\dfrac{11}{7}$ (E) $\dfrac{12}{7}$

19. Find the sum of the squares of all real numbers satisfying the equation

$$x^{256} - 256^{32} = 0.$$

(A) 8   (B) 128   (C) 512   (D) 65,536   (E) $2(256^{32})$

20. If $a = \tfrac{1}{2}$ and $(a + 1)(b + 1) = 2$, then the radian measure of Arctan a + Arctan b equals

(A) $\dfrac{\pi}{2}$ (B) $\dfrac{\pi}{3}$ (C) $\dfrac{\pi}{4}$ (D) $\dfrac{\pi}{5}$ (E) $\dfrac{\pi}{6}$

21. The length of the hypotenuse of a right triangle is h, and the radius of the inscribed circle is r. The ratio of the area of the circle to the area of the triangle is

(A) $\dfrac{\pi r}{h + 2r}$ (B) $\dfrac{\pi r}{h + r}$ (C) $\dfrac{\pi r}{2h + r}$ (D) $\dfrac{\pi r^2}{h^2 + r^2}$
(E) none of these

22. Find the number of pairs (m, n) of integers which satisfy the equation $m^3 + 6m^2 + 5m = 27n^3 + 9n^2 + 9n + 1$.

(A) 0   (B) 1   (C) 3   (D) 9   (E) infinitely many

23. The edges of a regular tetrahedron with vertices A, B, C and D each have length one. Find the least possible distance between a pair of points P and Q, where P is on edge AB and Q is on edge CD.

(A) $\dfrac{1}{2}$ (B) $\dfrac{3}{4}$

(C) $\dfrac{\sqrt{2}}{2}$ (D) $\dfrac{\sqrt{3}}{2}$

(E) $\dfrac{\sqrt{3}}{3}$

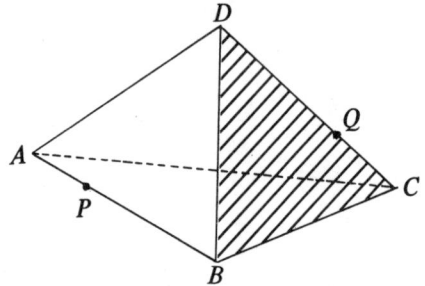

24. Sides AB, BC and CD of (simple†) quadrilateral $ABCD$ have lengths 4, 5 and 20, respectively. If vertex angles B and C are obtuse and $\sin C = -\cos B = \frac{3}{5}$, then side AD has length

(A) 24   (B) 24.5   (C) 24.6   (D) 24.8   (E) 25

25. If $q_1(x)$ and r_1 are the quotient and remainder, respectively, when the polynomial x^8 is divided by $x + \frac{1}{2}$, and if $q_2(x)$ and r_2 are the quotient and remainder, respectively, when $q_1(x)$ is divided by $x + \frac{1}{2}$, then r_2 equals

(A) $\dfrac{1}{256}$ (B) $-\dfrac{1}{16}$ (C) 1   (D) -16 (E) 256

26. The function f satisfies the functional equation

$$f(x) + f(y) = f(x + y) - xy - 1$$

for every pair x, y of real numbers. If $f(1) = 1$, then the number of integers $n \neq 1$ for which $f(n) = n$ is

(A) 0   (B) 1   (C) 2   (D) 3   (E) infinite

27. An ordered pair (b, c) of integers, each of which has absolute value less than or equal to five, is chosen at random, with each such ordered pair having an equal likelihood of being chosen. What is the probability that the equation $x^2 + bx + c = 0$ will *not* have distinct positive real roots?

(A) $\dfrac{106}{121}$ (B) $\dfrac{108}{121}$ (C) $\dfrac{110}{121}$ (D) $\dfrac{112}{121}$ (E) none of these

†A polygon is called "simple" if it is not self intersecting.

28. Circles with centers A, B and C each have radius r, where $1 < r < 2$. The distance between each pair of centers is 2. If B' is the point of intersection of circle A and circle C which is outside circle B, and if C' is the point of intersection of circle A and circle B which is outside circle C, then length $B'C'$ equals

(A) $3r - 2$

(B) r^2

(C) $r + \sqrt{3(r - 1)}$

(D) $1 + \sqrt{3(r^2 - 1)}$

(E) none of these

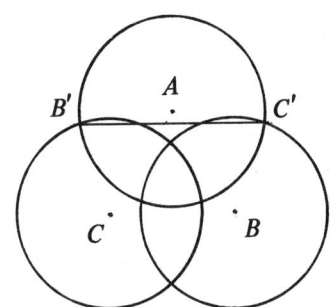

29. For each positive number x, let

$$f(x) = \frac{\left(x + \dfrac{1}{x}\right)^6 - \left(x^6 + \dfrac{1}{x^6}\right) - 2}{\left(x + \dfrac{1}{x}\right)^3 + \left(x^3 + \dfrac{1}{x^3}\right)}.$$

The minimum value of $f(x)$ is

(A) 1   (B) 2   (C) 3   (D) 4   (E) 6

30. In $\triangle ABC$, E is the midpoint of side BC and D is on side AC. If the length of AC is 1 and $\angle BAC = 60°$, $\angle ABC = 100°$, $\angle ACB = 20°$ and $\angle DEC = 80°$, then the area of $\triangle ABC$ plus twice the area of $\triangle CDE$ equals

(A) $\frac{1}{4}\cos 10°$ (B) $\frac{\sqrt{3}}{8}$

(C) $\frac{1}{4}\cos 40°$ (D) $\frac{1}{4}\cos 50°$

(E) $\frac{1}{8}$

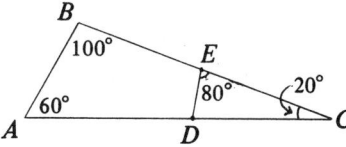

1980 Examination

1. The largest whole number such that seven times the number is less than 100 is

    (A) 12   (B) 13   (C) 14   (D) 15   (E) 16

2. The degree of $(x^2 + 1)^4(x^3 + 1)^3$ as a polynomial in x is

    (A) 5   (B) 7   (C) 12   (D) 17   (E) 72

3. If the ratio of $2x - y$ to $x + y$ is $\frac{2}{3}$, what is the ratio of x to y?

    (A) $\frac{1}{5}$ (B) $\frac{4}{5}$ (C) 1   (D) $\frac{6}{5}$ (E) $\frac{5}{4}$

4. In the adjoining figure, CDE is an equilateral triangle and $ABCD$ and $DEFG$ are squares. The measure of $\angle GDA$ is

    (A) 90°

    (B) 105°

    (C) 120°

    (D) 135°

    (E) 150°

    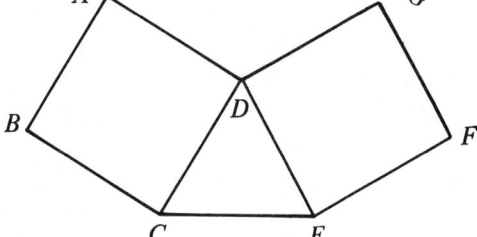

5. If AB and CD are perpendicular diameters of circle Q, and $\angle QPC = 60°$, then the length of PQ divided by the length of AQ is

    (A) $\frac{\sqrt{3}}{2}$ (B) $\frac{\sqrt{3}}{3}$

    (C) $\frac{\sqrt{2}}{2}$ (D) $\frac{1}{2}$

    (E) $\frac{2}{3}$

    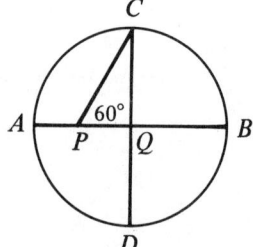

6. A positive number x satisfies the inequality $\sqrt{x} < 2x$ if and only if

(A) $x > \dfrac{1}{4}$ (B) $x > 2$ (C) $x > 4$ (D) $x < \dfrac{1}{4}$ (E) $x < 4$

7. Sides AB, BC, CD and DA of convex quadrilateral $ABCD$ have lengths 3, 4, 12, and 13, respectively; and $\angle CBA$ is a right angle. The area of the quadrilateral is

(A) 32 (B) 36
(C) 39 (D) 42
(E) 48

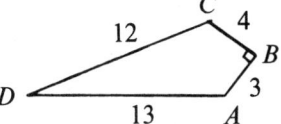

8. How many pairs (a, b) of non-zero real numbers satisfy the equation

$$\frac{1}{a} + \frac{1}{b} = \frac{1}{a + b}?$$

(A) none (B) 1 (C) 2 (D) one pair for each $b \neq 0$
(E) two pairs for each $b \neq 0$

9. A man walks x miles due west, turns $150°$ to his left and walks 3 miles in the new direction. If he finishes at a point $\sqrt{3}$ miles from his starting point, then x is

(A) $\sqrt{3}$ (B) $2\sqrt{3}$ (C) $\dfrac{3}{2}$ (D) 3

(E) not uniquely determined by the given information

10. The number of teeth in three meshed circular gears A, B, C are x, y, z, respectively. (The teeth on all gears are the same size and regularly spaced as in the figure.) The angular speeds, in revolutions per minute, of A, B, C are in the proportion

(A) $x:y:z$ (B) $z:y:x$ (C) $y:z:x$ (D) $yz:xz:xy$
(E) $xz:yx:zy$

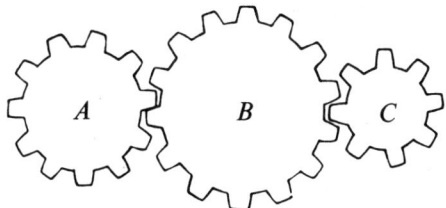

11. If the sum of the first 10 terms and the sum of the first 100 terms of a given arithmetic progression are 100 and 10, respectively, then the sum of the first 110 terms is

(A) 90   (B) -90 (C) 110   (D) -110 (E) -100

12. The equations of L_1 and L_2 are $y = mx$ and $y = nx$, respectively. Suppose L_1 makes twice as large an angle with the horizontal (measured counterclockwise from the positive x-axis) as does L_2, and that L_1 has 4 times the slope of L_2. If L_1 is not horizontal, then mn is

(A) $\dfrac{\sqrt{2}}{2}$ (B) $-\dfrac{\sqrt{2}}{2}$ (C) 2   (D) -2

(E) not uniquely determined by the given information

13. A bug (of negligible size) starts at the origin on the co-ordinate plane. First it moves 1 unit right to $(1, 0)$. Then it makes a $90°$ turn counterclockwise and travels $\frac{1}{2}$ a unit to $\left(1, \frac{1}{2}\right)$. If it continues in this fashion, each time making a $90°$ turn counterclockwise and traveling half as far as in the previous move, to which of the following points will it come closest?

(A) $\left(\dfrac{2}{3}, \dfrac{2}{3}\right)$ (B) $\left(\dfrac{4}{5}, \dfrac{2}{5}\right)$ (C) $\left(\dfrac{2}{3}, \dfrac{4}{5}\right)$ (D) $\left(\dfrac{2}{3}, \dfrac{1}{3}\right)$

(E) $\left(\dfrac{2}{5}, \dfrac{4}{5}\right)$

14. If the function f defined by

$$f(x) = \frac{cx}{2x + 3}, \qquad x \neq -\frac{3}{2}, \qquad c \text{ a constant,}$$

satisfies $f(f(x)) = x$ for all real numbers x except $-\frac{3}{2}$, then c is

(A) -3 (B) $-\dfrac{3}{2}$ (C) $\dfrac{3}{2}$ (D) 3

(E) not uniquely determined by the given information

15. A store prices an item in dollars and cents so that when 4% sales tax is added no rounding is necessary because the result is exactly n dollars, where n is a positive integer. The smallest value of n is

(A) 1   (B) 13   (C) 25   (D) 26   (E) 100

16. Four of the eight vertices of a cube are vertices of a regular tetrahedron. Find the ratio of the surface area of the cube to the surface area of the tetrahedron.

(A) $\sqrt{2}$ (B) $\sqrt{3}$

(C) $\sqrt{\dfrac{3}{2}}$ (D) $\dfrac{2}{\sqrt{3}}$

(E) 2

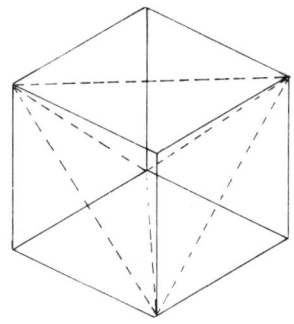

17. Given that $i^2 = -1$, for how many integers n is $(n + i)^4$ an integer?

(A) none   (B) 1   (C) 2   (D) 3   (E) 4

18. If $b > 1$, $\sin x > 0$, $\cos x > 0$ and $\log_b \sin x = a$, then $\log_b \cos x$ equals

(A) $2 \log_b (1 - b^{a/2})$ (B) $\sqrt{1 - a^2}$ (C) b^{a^2}

(D) $\dfrac{1}{2} \log_b (1 - b^{2a})$ (E) none of these

19. Let C_1, C_2 and C_3 be three parallel chords of a circle on the same side of the center. The distance between C_1 and C_2 is the same as the distance between C_2 and C_3. The lengths of the chords are 20, 16 and 8. The radius of the circle is

(A) 12   (B) $4\sqrt{7}$ (C) $\dfrac{5\sqrt{65}}{3}$ (D) $\dfrac{5\sqrt{22}}{2}$

(E) not uniquely determined by the given information

20. A box contains 2 pennies, 4 nickels and 6 dimes. Six coins are drawn without replacement, with each coin having an equal probability of being chosen. What is the probability that the value of the coins drawn is at least 50 cents?

(A) $\dfrac{37}{924}$ (B) $\dfrac{91}{924}$ (C) $\dfrac{127}{924}$ (D) $\dfrac{132}{924}$

(E) none of these

21. In triangle ABC, $\angle CBA = 72°$, E is the midpoint of side AC, and D is a point on side BC such that $2BD = DC$; AD and BE intersect at F. The ratio of the area of $\triangle BDF$ to the area of quadrilateral $FDCE$ is

(A) $\dfrac{1}{5}$ (B) $\dfrac{1}{4}$

(C) $\dfrac{1}{3}$ (D) $\dfrac{2}{5}$

(E) none of these

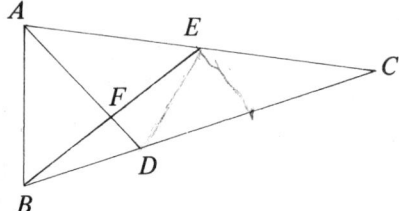

22. For each real number x, let $f(x)$ be the minimum of the numbers $4x + 1$, $x + 2$, and $-2x + 4$. Then the maximum value of $f(x)$ is

(A) $\dfrac{1}{3}$ (B) $\dfrac{1}{2}$ (C) $\dfrac{2}{3}$ (D) $\dfrac{5}{2}$ (E) $\dfrac{8}{3}$

23. Line segments drawn from the vertex opposite the hypotenuse of a right triangle to the points trisecting the hypotenuse have lengths $\sin x$ and $\cos x$, where x is a real number such that $0 < x < \dfrac{\pi}{2}$. The length of the hypotenuse is

(A) $\dfrac{4}{3}$ (B) $\dfrac{3}{2}$ (C) $\dfrac{3\sqrt{5}}{5}$ (D) $\dfrac{2\sqrt{5}}{3}$

(E) not uniquely determined by the given information

24. For some real number r, the polynomial $8x^3 - 4x^2 - 42x + 45$ is divisible by $(x - r)^2$. Which of the following numbers is closest to r?

(A) 1.22   (B) 1.32   (C) 1.42   (D) 1.52   (E) 1.62

25. In the non-decreasing sequence of odd integers $\{a_1, a_2, a_3, \ldots\} = \{1, 3, 3, 3, 5, 5, 5, 5, 5, \ldots\}$ each positive odd integer k appears k times. It is a fact that there are integers b, c and d such that, for all positive integers n,

$$a_n = b\left[\sqrt{n + c}\,\right] + d,$$

where $[x]$ denotes the largest integer not exceeding x. The sum $b + c + d$ equals

(A) 0   (B) 1   (C) 2   (D) 3   (E) 4

26. Four balls of radius 1 are mutually tangent, three resting on the floor and the fourth resting on the others. A tetrahedron, each of whose edges has length s, is circumscribed around the balls. Then s equals

(A) $4\sqrt{2}$ (B) $4\sqrt{3}$ (C) $2\sqrt{6}$ (D) $1 + 2\sqrt{6}$ (E) $2 + 2\sqrt{6}$

27. The sum $\sqrt[3]{5 + 2\sqrt{13}} + \sqrt[3]{5 - 2\sqrt{13}}$ equals

(A) $\dfrac{3}{2}$ (B) $\dfrac{\sqrt[3]{65}}{4}$ (C) $\dfrac{1 + \sqrt[6]{13}}{2}$ (D) $\sqrt[3]{2}$

(E) none of these

28. The polynomial $x^{2n} + 1 + (x + 1)^{2n}$ is not divisible by $x^2 + x + 1$ if n equals

(A) 17 (B) 20 (C) 21 (D) 64 (E) 65

29. How many ordered triples (x, y, z) of integers satisfy the system of equations below?

$$
\begin{aligned}
x^2 - 3xy + 2y^2 \quad\quad\; - z^2 &= 31, \\
-x^2 \quad\quad\quad\; + 6yz + 2z^2 &= 44, \\
x^2 + \; xy \quad\quad\quad\; + 8z^2 &= 100.
\end{aligned}
$$

(A) 0 (B) 1 (C) 2 (D) a finite number greater than two
(E) infinitely many

30. A six digit number (base 10) is *squarish* if it satisfies the following conditions:

(i) none of its digits is zero;
(ii) it is a perfect square; and
(iii) the first two digits, the middle two digits and the last two digits of the number are all perfect squares when considered as two digit numbers.

How many squarish numbers are there?

(A) 0 (B) 2 (C) 3 (D) 8 (E) 9

1981 Examination

1. If $\sqrt{x+2} = 2$ then $(x+2)^2$ equals

   (A) $\sqrt{2}$ (B) 2   (C) 4   (D) 8   (E) 16

2. Point E is on side AB of square $ABCD$. If EB has length one and EC has length two, then the area of the square is

   (A) $\sqrt{3}$ (B) $\sqrt{5}$

   (C) 3   (D) $2\sqrt{3}$

   (E) 5

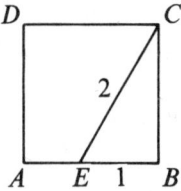

3. For $x \neq 0$, $\dfrac{1}{x} + \dfrac{1}{2x} + \dfrac{1}{3x}$ equals

   (A) $\dfrac{1}{2x}$ (B) $\dfrac{1}{6x}$ (C) $\dfrac{5}{6x}$ (D) $\dfrac{11}{6x}$ (E) $\dfrac{1}{6x^3}$

4. If three times the larger of two numbers is four times the smaller and the difference between the numbers is 8, then the larger of the two numbers is

   (A) 16   (B) 24   (C) 32   (D) 44   (E) 52

5. In trapezoid $ABCD$, sides AB and CD are parallel, and diagonal BD and side AD have equal length. If $\angle DCB = 110°$ and $\angle CBD = 30°$, then $\angle ADB =$

   (A) 80°   (B) 90°

   (C) 100°   (D) 110°

   (E) 120°

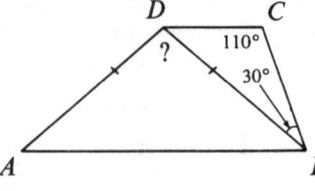

6. If $\dfrac{x}{x-1} = \dfrac{y^2 + 2y - 1}{y^2 + 2y - 2}$, then x equals

   (A) $y^2 + 2y - 1$ (B) $y^2 + 2y - 2$ (C) $y^2 + 2y + 2$
   (D) $y^2 + 2y + 1$ (E) $-y^2 - 2y + 1$

7. How many of the first one hundred positive integers are divisible by all of the numbers 2, 3, 4, 5?

   (A) 0   (B) 1   (C) 2   (D) 3   (E) 4

8. For all positive numbers x, y, z, the product

   $$(x+y+z)^{-1}(x^{-1}+y^{-1}+z^{-1})(xy+yz+zx)^{-1}\left[(xy)^{-1}+(yz)^{-1}+(zx)^{-1}\right]$$

   equals

   (A) $x^{-2}y^{-2}z^{-2}$ (B) $x^{-2}+y^{-2}+z^{-2}$ (C) $(x+y+z)^{-2}$

   (D) $\dfrac{1}{xyz}$ (E) $\dfrac{1}{xy+yz+zx}$

9. In the adjoining figure, PQ is a diagonal of the cube. If PQ has length a, then the surface area of the cube is

   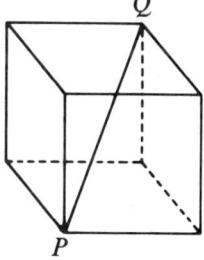

   (A) $2a^2$ (B) $2\sqrt{2}\,a^2$
   (C) $2\sqrt{3}\,a^2$ (D) $3\sqrt{3}\,a^2$
   (E) $6a^2$

10. The lines L and K are symmetric to each other with respect to the line $y = x$. If the equation of line L is $y = ax + b$ with $a \neq 0$ and $b \neq 0$, then the equation of K is $y =$

    (A) $\dfrac{1}{a}x + b$ (B) $-\dfrac{1}{a}x + b$ (C) $-\dfrac{1}{a}x - \dfrac{b}{a}$

    (D) $\dfrac{1}{a}x + \dfrac{b}{a}$ (E) $\dfrac{1}{a}x - \dfrac{b}{a}$

11. The three sides of a right triangle have integral lengths which form an arithmetic progression. One of the sides could have length

    (A) 22   (B) 58   (C) 81   (D) 91   (E) 361

12. If p, q and M are positive numbers and $q < 100$, then the number obtained by increasing M by $p\%$ and decreasing the result by $q\%$ exceeds M if and only if

    (A) $p > q$ (B) $p > \dfrac{q}{100 - q}$ (C) $p > \dfrac{q}{1 - q}$

    (D) $p > \dfrac{100q}{100 + q}$ (E) $p > \dfrac{100q}{100 - q}$

13. Suppose that at the end of any year, a unit of money has lost 10% of the value it had at the beginning of that year. Find the smallest integer n such that after n years the unit of money will have lost at least 90% of its value. (To the nearest thousandth $\log_{10} 3$ is .477.)

(A) 14   (B) 16   (C) 18   (D) 20   (E) 22

14. In a geometric sequence of real numbers, the sum of the first two terms is 7, and the sum of the first six terms is 91. The sum of the first four terms is

(A) 28   (B) 32   (C) 35   (D) 49   (E) 84

15. If $b > 1$, $x > 0$ and $(2x)^{\log_b 2} - (3x)^{\log_b 3} = 0$,   then x is

(A) $\dfrac{1}{216}$ (B) $\dfrac{1}{6}$ (C) 1   (D) 6

(E) not uniquely determined

16. The base three representation of x is
$$121122111122211112222.$$
The first digit (on the left) of the base nine representation of x is

(A) 1   (B) 2   (C) 3   (D) 4   (E) 5

17. The function f is not defined for $x = 0$, but, for all non-zero real numbers x,

$$f(x) + 2f\left(\frac{1}{x}\right) = 3x.$$ The equation $f(x) = f(-x)$ is satisfied by

(A) exactly one real number
(B) exactly two real numbers
(C) no real numbers
(D) infinitely many, but not all, non-zero real numbers
(E) all non-zero real numbers

18. The number of real solutions to the equation

$$\frac{x}{100} = \sin x$$

is

(A) 61   (B) 62   (C) 63   (D) 64   (E) 65

19. In $\triangle ABC$, M is the midpoint of side BC, AN bisects $\measuredangle BAC$, $BN \perp AN$ and θ is the measure of $\measuredangle BAC$. If sides AB and AC have lengths 14 and 19, respectively, then length MN equals

(A) 2   (B) $\dfrac{5}{2}$

(C) $\dfrac{5}{2} - \sin\theta$

(D) $\dfrac{5}{2} - \dfrac{1}{2}\sin\theta$

(E) $\dfrac{5}{2} - \dfrac{1}{2}\sin\left(\dfrac{\theta}{2}\right)$

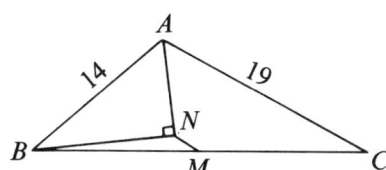

20. A ray of light originates from point A and travels in a plane, being reflected n times between lines AD and CD, before striking a point B (which may be on AD or CD) perpendicularly and retracing its path to A. (At each point of reflection the light makes two equal angles as indicated in the adjoining figure. The figure shows the light path for $n = 3$.) If $\measuredangle CDA = 8°$, what is the largest value n can have?

(A) 6   (B) 10   (C) 38   (D) 98
(E) There is no largest value.

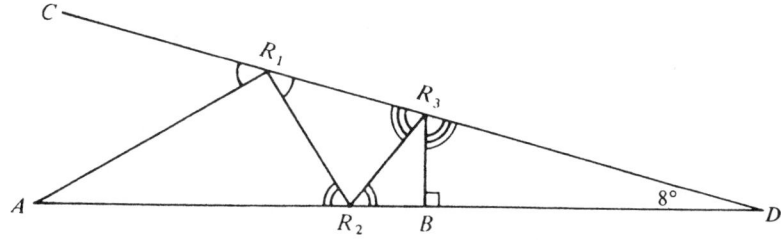

21. In a triangle with sides of lengths a, b and c,

$$(a + b + c)(a + b - c) = 3ab.$$

The measure of the angle opposite the side of length c is

(A) 15°   (B) 30°   (C) 45°   (D) 60°   (E) 150°

22. How many lines in a three-dimensional rectangular coordinate system pass through four distinct points of the form (i, j, k), where i, j and k are positive integers not exceeding four?

(A) 60   (B) 64   (C) 72   (D) 76   (E) 100

23. Equilateral $\triangle ABC$ is inscribed in a circle. A second circle is tangent internally to the circumcircle at T and tangent to sides AB and AC at points P and Q. If side BC has length 12, then segment PQ has length

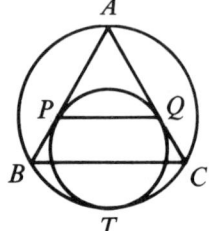

(A) 6

(B) $6\sqrt{3}$

(C) 8

(D) $8\sqrt{3}$

(E) 9

24. If θ is a constant such that $0 < \theta < \pi$ and $x + \dfrac{1}{x} = 2\cos\theta$, then for each positive integer n, $x^n + \dfrac{1}{x^n}$ equals

(A) $2\cos\theta$ (B) $2^n\cos\theta$ (C) $2\cos^n\theta$

(D) $2\cos n\theta$ (E) $2^n\cos^n\theta$

25. In triangle ABC in the adjoining figure, AD and AE trisect $\angle BAC$. The lengths of BD, DE and EC are 2, 3, and 6, respectively. The length of the shortest side of $\triangle ABC$ is

(A) $2\sqrt{10}$

(B) 11

(C) $6\sqrt{6}$

(D) 6

(E) not uniquely determined by the given information

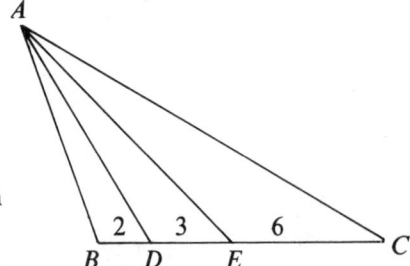

26. Alice, Bob and Carol repeatedly take turns tossing a die. Alice begins; Bob always follows Alice; Carol always follows Bob; and Alice always follows Carol. Find the probability that Carol will be the first one to toss a six. (The probability of obtaining a six on any toss is $\frac{1}{6}$, independent of the outcome of any other toss.)

(A) $\dfrac{1}{3}$ (B) $\dfrac{2}{9}$ (C) $\dfrac{5}{18}$ (D) $\dfrac{25}{91}$ (E) $\dfrac{36}{91}$

27. In the adjoining figure triangle ABC is inscribed in a circle. Point D lies on $\overset{\frown}{AC}$ with $\overset{\frown}{DC} = 30°$, and point G lies on $\overset{\frown}{BA}$ with $\overset{\frown}{BG} > \overset{\frown}{GA}$. Side AB and side AC each have length equal to the length of chord DG, and $\angle CAB = 30°$. Chord DG intersects sides AC and AB at E and F, respectively. The ratio of the area of $\triangle AFE$ to the area of $\triangle ABC$ is

(A) $\dfrac{2 - \sqrt{3}}{3}$ (B) $\dfrac{2\sqrt{3} - 3}{3}$

(C) $7\sqrt{3} - 12$ (D) $3\sqrt{3} - 5$

(E) $\dfrac{9 - 5\sqrt{3}}{3}$

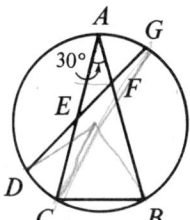

28. Consider the set of all equations $x^3 + a_2 x^2 + a_1 x + a_0 = 0$, where a_2, a_1, a_0 are real constants and $|a_i| \leq 2$ for $i = 0, 1, 2$. Let r be the largest positive real number which satisfies at least one of these equations. Then

(A) $1 \leq r < \dfrac{3}{2}$ (B) $\dfrac{3}{2} \leq r < 2$ (C) $2 \leq r < \dfrac{5}{2}$

(D) $\dfrac{5}{2} \leq r < 3$ (E) $3 \leq r < \dfrac{7}{2}$

29. If $a \geq 1$, then the sum of the real solutions of

$$\sqrt{a - \sqrt{a + x}} = x$$

is equal to

(A) $\sqrt{a} - 1$ (B) $\dfrac{\sqrt{a} - 1}{2}$ (C) $\sqrt{a - 1}$

(D) $\dfrac{\sqrt{a} - 1}{2}$ (E) $\dfrac{\sqrt{4a - 3} - 1}{2}$

30. If a, b, c, d are the solutions of the equation $x^4 - bx - 3 = 0$, then an equation whose solutions are

$$\dfrac{a + b + c}{d^2}, \quad \dfrac{a + b + d}{c^2}, \quad \dfrac{a + c + d}{b^2}, \quad \dfrac{b + c + d}{a^2}$$

is

(A) $3x^4 + bx + 1 = 0$ (B) $3x^4 - bx + 1 = 0$
(C) $3x^4 + bx^3 - 1 = 0$ (D) $3x^4 - bx^3 - 1 = 0$
(E) none of these

1982 Examination

1. When the polynomial $x^3 - 2$ is divided by the polynomial $x^2 - 2$, the remainder is

    (A) 2   (B) -2 (C) $-2x - 2$ (D) $2x + 2$ (E) $2x - 2$

2. If a number eight times as large as x is increased by two, then one fourth of the result equals

    (A) $2x + \dfrac{1}{2}$ (B) $x + \dfrac{1}{2}$ (C) $2x + 2$ (D) $2x + 4$

    (E) $2x + 16$

3. Evaluate $(x^x)^{(x^x)}$ at $x = 2$.

    (A) 16   (B) 64   (C) 256   (D) 1024   (E) 65,536

4. The perimeter of a semicircular region, measured in centimeters, is numerically equal to its area, measured in square centimeters. The radius of the semicircle, measured in centimeters, is

    (A) π (B) $\dfrac{2}{\pi}$ (C) 1   (D) $\dfrac{1}{2}$ (E) $\dfrac{4}{\pi} + 2$

5. Two positive numbers x and y are in the ratio $a : b$, where $0 < a < b$. If $x + y = c$, then the smaller of x and y is

    (A) $\dfrac{ac}{b}$ (B) $\dfrac{bc - ac}{b}$ (C) $\dfrac{ac}{a + b}$ (D) $\dfrac{bc}{a + b}$

    (E) $\dfrac{ac}{b - a}$

6. The sum of all but one of the interior angles of a convex polygon equals $2570°$. The remaining angle is

    (A) $90°$ (B) $105°$ (C) $120°$ (D) $130°$ (E) $144°$

7. If the operation $x * y$ is defined by $x * y = (x + 1)(y + 1) - 1$, then which one of the following is **false**?

    (A) $x * y = y * x$ for all real x and y.
    (B) $x * (y + z) = (x * y) + (x * z)$ for all real x, y, and z.
    (C) $(x - 1) * (x + 1) = (x * x) - 1$ for all real x.
    (D) $x * 0 = x$ for all real x.
    (E) $x * (y * z) = (x * y) * z$ for all real x, y, and z.

8. By definition $r! = r(r-1) \cdots 1$ and
$$\binom{j}{k} = \frac{j!}{k!(j-k)!},$$
where r, j, k are positive integers and $k < j$. If $\binom{n}{1}, \binom{n}{2}, \binom{n}{3}$ form an arithmetic progression with $n > 3$, then n equals

(A) 5 (B) 7 (C) 9 (D) 11 (E) 12

9. A vertical line divides the triangle with vertices $(0,0)$, $(1,1)$ and $(9,1)$ in the xy-plane into two regions of equal area. The equation of the line is $x =$

(A) 2.5 (B) 3.0 (C) 3.5 (D) 4.0 (E) 4.5

10. In the adjoining diagram, BO bisects $\angle CBA$, CO bisects $\angle ACB$, and MN is parallel to BC. If $AB = 12$, $BC = 24$, and $AC = 18$, then the perimeter of $\triangle AMN$ is

(A) 30 (B) 33
(C) 36 (D) 39
(E) 42

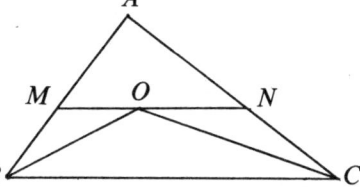

11. How many integers with four different digits are there between 1,000 and 9,999 such that the absolute value of the difference between the first digit and the last digit is 2?

(A) 672 (B) 784 (C) 840 (D) 896 (E) 1,008

12. Let $f(x) = ax^7 + bx^3 + cx - 5$, where a, b and c are constants. If $f(-7) = 7$, then $f(7)$ equals

(A) -17 (B) -7 (C) 14 (D) 21
(E) not uniquely determined

13. If $a > 1$, $b > 1$ and $p = \dfrac{\log_b(\log_b a)}{\log_b a}$, then a^p equals

(A) 1 (B) b (C) $\log_a b$ (D) $\log_b a$ (E) $a^{\log_b a}$

14. In the adjoining figure, points B and C lie on line segment AD, and AB, BC and CD are diameters of circles O, N and P, respectively. Circles O, N and P all have radius 15, and the line AG is tangent to circle P at G. If AG intersects circle N at points E and F, then chord EF has length

(A) 20   (B) $15\sqrt{2}$
(C) 24   (D) 25
(E) none of these

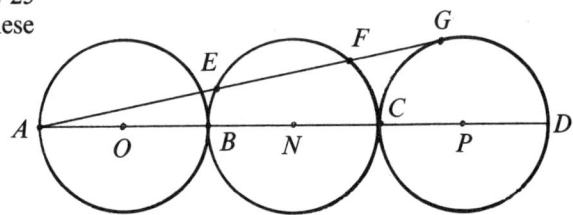

15. Let $[z]$ denote the greatest integer not exceeding z. Let x and y satisfy the simultaneous equations

$$y = 2[x] + 3$$
$$y = 3[x - 2] + 5.$$

If x is not an integer, then $x + y$ is

(A) an integer   (B) between 4 and 5
(C) between -4 and 4   (D) between 15 and 16
(E) 16.5

16. In the adjoining figure, a wooden cube has edges of length 3 meters. Square holes of side one meter, centered in each face, are cut through to the opposite face. The edges of the holes are parallel to the edges of the cube. The entire surface area including the inside, in square meters, is

(A) 54   (B) 72

(C) 76   (D) 84

(E) 86

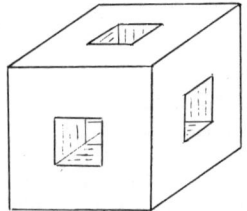

17. How many real numbers x satisfy the equation

$$3^{2x+2} - 3^{x+3} - 3^x + 3 = 0 ?$$

(A) 0   (B) 1   (C) 2   (D) 3   (E) 4

18. In the adjoining figure of a rectangular solid, $\measuredangle DHG = 45°$ and $\measuredangle FHB = 60°$. Find the cosine of $\measuredangle BHD$.

(A) $\dfrac{\sqrt{3}}{6}$ (B) $\dfrac{\sqrt{2}}{6}$ (C) $\dfrac{\sqrt{6}}{3}$ (D) $\dfrac{\sqrt{6}}{4}$ (E) $\dfrac{\sqrt{6} - \sqrt{2}}{4}$

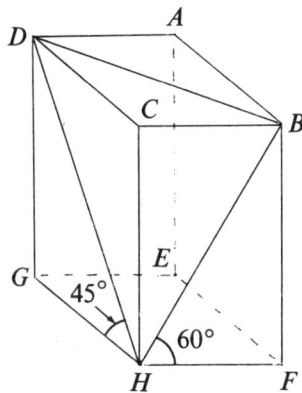

19. Let $f(x) = |x - 2| + |x - 4| - |2x - 6|$, for $2 \leqslant x \leqslant 8$. The sum of the largest and smallest values of $f(x)$ is

(A) 1   (B) 2   (C) 4   (D) 6   (E) none of these

20. The number of pairs of positive integers (x, y) which satisfy the equation $x^2 + y^2 = x^3$ is

(A) 0   (B) 1   (C) 2   (D) not finite   (E) none of these

21. In the adjoining figure, the triangle ABC is a right triangle with $\measuredangle BCA = 90°$. Median CM is perpendicular to median BN, and side $BC = s$. The length of BN is

(A) $s\sqrt{2}$ (B) $\dfrac{3}{2}s\sqrt{2}$ (C) $2s\sqrt{2}$ (D) $\dfrac{s\sqrt{5}}{2}$ (E) $\dfrac{s\sqrt{6}}{2}$

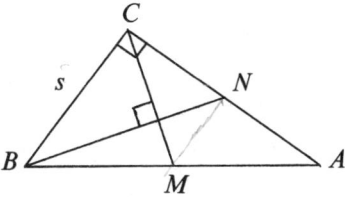

22. In a narrow alley of width w a ladder of length a is placed with its foot at a point P between the walls. Resting against one wall at Q, a distance k above the ground, the ladder makes a 45° angle with the ground. Resting against the other wall at R, a distance h above the ground, the ladder makes a 75° angle with the ground. The width w is equal to

(A) a (B) RQ

(C) k (D) $\dfrac{h+k}{2}$

(E) h

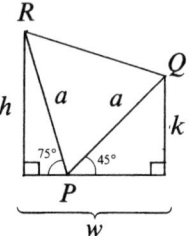

23. The lengths of the sides of a triangle are consecutive integers, and the largest angle is twice the smallest angle. The cosine of the smallest angle is

(A) $\dfrac{3}{4}$ (B) $\dfrac{7}{10}$ (C) $\dfrac{2}{3}$ (D) $\dfrac{9}{14}$ (E) none of these

24. In the adjoining figure, the circle meets the sides of an equilateral triangle at six points. If $AG = 2$, $GF = 13$, $FC = 1$ and $HJ = 7$, then DE equals

(A) $2\sqrt{22}$ (B) $7\sqrt{3}$

(C) 9   (D) 10

(E) 13

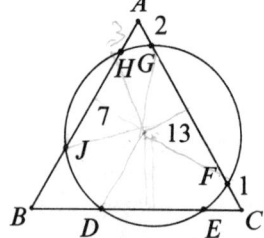

25. The adjoining figure is a map of part of a city: the small rectangles are blocks and the spaces in between are streets. Each morning a student walks from intersection A to intersection B, always walking along streets shown, always going east or south. For variety, at each intersection where he has a choice, he chooses with probability $\frac{1}{2}$ (independent of all other choices) whether to go east or south. Find the probability that, on any given morning, he walks through intersection C.

(A) $\dfrac{11}{32}$ (B) $\dfrac{1}{2}$

(C) $\dfrac{4}{7}$ (D) $\dfrac{21}{32}$

(E) $\dfrac{3}{4}$

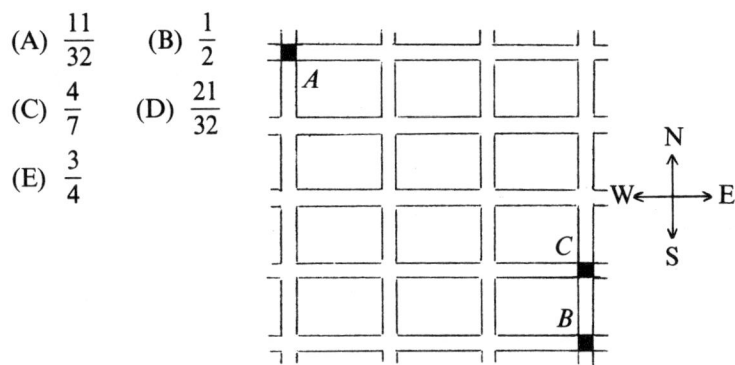

26. If the base 8 representation of a perfect square is $ab3c$, where $a \neq 0$, then c is

    (A) (0)   (B) 1   (C) 3   (D) 4   (E) not uniquely determined

27. Suppose $z = a + bi$ is a solution of the polynomial equation

    $$c_4 z^4 + i c_3 z^3 + c_2 z^2 + i c_1 z + c_0 = 0,$$

    where $c_0, c_1, c_2, c_3, c_4, a$ and b are real constants and $i^2 = -1$. Which one of the following must also be a solution?

    (A) $-a - bi$ (B) $a - bi$ (C) $-a + bi$ (D) $b + ai$
    (E) none of these

28. A set of consecutive positive integers beginning with 1 is written on a blackboard. One number is erased. The average (arithmetic mean) of the remaining numbers is $35\frac{7}{17}$. What number was erased?

    (A) 6   (B) 7   (C) 8   (D) 9   (E) can not be determined

29. Let x, y and z be three positive real numbers whose sum is 1. If no one of these numbers is more than twice any other, then the minimum possible value of the product xyz is

    (A) $\dfrac{1}{32}$ (B) $\dfrac{1}{36}$ (C) $\dfrac{4}{125}$ (D) $\dfrac{1}{127}$ (E) none of these

30. Find the units digit in the decimal expansion of

    $$\left(15 + \sqrt{220}\right)^{19} + \left(15 + \sqrt{220}\right)^{82}.$$

    (A) 0   (B) 2   (C) 5   (D) 9   (E) none of these

II

Answer Keys

1973 Answers

1. D	8. E	15. D	22. A	29. A
2. C	9. E	16. B	23. D	30. B
3. B	10. A	17. E	24. D	31. C
4. D	11. B	18. C	25. E	32. A
5. D	12. D	19. D	26. E	33. C
6. C	13. D	20. C	27. A	34. C
7. A	14. C	21. B	28. C	35. E

1974 Answers

1. D	7. D	13. D	19. A	25. C
2. B	8. A	14. A	20. D	26. C
3. A	9. B	15. B	21. B	27. A
4. D	10. B	16. A	22. E	28. D
5. B	11. A	17. C	23. B	29. B
6. D	12. B	18. D	24. A	30. A

1975 Answers

1. B	7. E	13. D	19. D	25. B
2. D	8. D	14. E	20. B	26. C
3. A	9. C	15. A	21. D	27. E
4. A	10. A	16. C	22. E	28. A
5. B	11. E	17. D	23. C	29. C
6. E	12. B	18. D	24. E	30. B

1976 Answers

1. B	7. E	13. A	19. B	25. D
2. B	8. A	14. A	20. E	26. C
3. E	9. D	15. B	21. B	27. A
4. C	10. D	16. E	22. A	28. B
5. C	11. B	17. A	23. A	29. B
6. C	12. C	18. E	24. C	30. E

1977 Answers

1. D	7. E	13. E	19. A	25. E
2. D	8. B	14. B	20. E	26. B
3. E	9. B	15. D	21. B	27. C
4. C	10. E	16. D	22. C	28. A
5. A	11. B	17. B	23. B	29. B
6. D	12. D	18. B	24. D	30. A

1978 Answers

1. B	7. E	13. B	19. C	25. D
2. C	8. D	14. C	20. A	26. B
3. D	9. B	15. A	21. A	27. C
4. B	10. B	16. E	22. D	28. E
5. C	11. C	17. D	23. C	29. D
6. E	12. E	18. C	24. A	30. E

1979 Answers

1. D	7. E	13. A	19. A	25. B
2. D	8. C	14. C	20. C	26. B
3. C	9. E	15. E	21. B	27. E
4. E	10. D	16. E	22. A	28. D
5. D	11. B	17. C	23. C	29. E
6. A	12. B	18. C	24. E	30. B

1980 Answers

1. C	7. B	13. B	19. D	25. C
2. D	8. A	14. A	20. C	26. E
3. E	9. E	15. B	21. A	27. E
4. C	10. D	16. B	22. E	28. C
5. B	11. D	17. D	23. C	29. A
6. A	12. C	18. D	24. D	30. B

1981 Answers

1. E	7. B	13. E	19. B	25. A
2. C	8. A	14. A	20. B	26. D
3. D	9. A	15. B	21. D	27. C
4. C	10. E	16. E	22. D	28. D
5. C	11. C	17. B	23. C	29. E
6. A	12. E	18. C	24. D	30. D

1982 Answers

1. E	7. B	13. D	19. B	25. D
2. A	8. B	14. C	20. D	26. B
3. C	9. B	15. D	21. E	27. C
4. E	10. A	16. B	22. E	28. B
5. C	11. C	17. C	23. A	29. A
6. D	12. A	18. D	24. A	30. D

III

Solutions†

1973 Solutions

Part 1

1. (D) Let O denote the center of the circle, and let OR and AB be the radius and the chord which are perpendicular bisectors of each other at M. Applying the Pythagorean theorem to right triangle OMA yields

$$(AM)^2 = (OA)^2 - (OM)^2$$
$$= 12^2 - 6^2 = 108,$$
$$AM = 6\sqrt{3}.$$

Thus the required chord has length $12\sqrt{3}$.

2. (C) The unpainted cubes form the $8 \times 8 \times 8$ cube of interior cubes. Therefore, $10^3 - 8^3 = 488$ cubes have at least one face painted.

---

† The letter following the problem number refers to the correct choice of the five listed in the examination.

3. (B) Thirteen is the smallest prime p such that $126 - p$ is also prime. Thus the largest difference is $113 - 13 = 100$.

4. (D) In the adjoining figure MV is an altitude of $\triangle ABV$. Since $\triangle AMV$ is a $30° - 60° - 90°$ triangle, MV has length $2\sqrt{3}$. The required area is, therefore, area $\triangle ABV = \frac{1}{2}(AB)(MV)$ $= \frac{1}{2}(12)2\sqrt{3} = 12\sqrt{3}$.

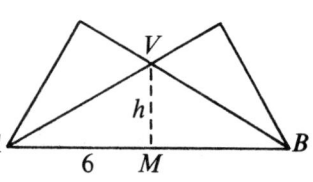

5. (D) Let $a * b$ denote the average, $\frac{1}{2}(a + b)$, of a and b. Then

II $\qquad a * b = \frac{1}{2}(a + b) = \frac{1}{2}(b + a) = b * a,$

and

IV $\quad a + (b * c) = a + \frac{1}{2}(b + c) = \frac{1}{2}(a + b + a + c)$
$$= (a + b) * (a + c),$$

but

I $\qquad (a * b) * c = \frac{1}{2}[\frac{1}{2}(a + b) + c] = \frac{1}{4}a + \frac{1}{4}b + \frac{1}{2}c,$

while

$$a * (b * c) = \frac{1}{2}[a + \frac{1}{2}(b + c)] = \frac{1}{2}a + \frac{1}{4}b + \frac{1}{4}c;$$

and

III $\qquad a * (b + c) = \frac{1}{2}(a + b + c),$

while

$$(a * b) + (a * c) = \frac{1}{2}(a + b) + \frac{1}{2}(a + c) \neq a * (b + c).$$

To see that V is false, suppose that $e * a = a$ for some e and all a. Then $\frac{1}{2}(e + a) = a$, so $e = a$. Clearly this cannot hold for more than one value of a. Thus only II and IV are true.

6. (C) Let $b > 5$ be the base. Since $(24_b)^2 = (2b + 4)^2 = 4b^2 + 16b + 16$ and $554_b = 5b^2 + 5b + 4$, it follows that

$$5b^2 + 5b + 4 = 4b^2 + 16b + 16,$$

so

$$b^2 - 11b - 12 = (b + 1)(b - 12) = 0,$$

$b = -1$ or $b = 12$.

7. (A) The numbers to be added form an arithmetic progression with first term $a = 51$, last term $l = 341$ and common difference $d = 10$. We use the formula $l = a + (n - 1)d$ to determine the number n of terms to be summed and $S = \frac{1}{2}n(a + l)$ to find their sum. Thus $341 = 51 + (n - 1)10$ yields $n = 30$, so $S = 15(392) = 5880$.

8. (E) The number of pints P of paint needed is proportional to the surface area (hence to the square of the height h) of a statue and also to the number n of statues to be painted. Thus $P = knh^2$; substituting $n = 1$, $h = 6$ and $P = 1$ into this equation, we obtain the constant of proportionality $k = 1/36$, so that

$$P = \frac{1}{36}nh^2.$$

Now when $n = 540$ and $h = 1$, $P = (1/36)(540)(1^2) = 15$.

9. (E) Right triangles CHM and CHB are congruent since their angles at C are equal. Therefore the base MH of $\triangle CMH$ is $\frac{1}{4}$ of the base AB of $\triangle ABC$, while their altitudes are equal. Hence the area of $\triangle ABC$ is $4K$.

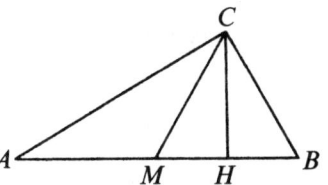

10. (A) If $n \neq -1$, then elementary operations on the system yield the solution $x = y = z = 1/(n + 1)$. If $n = -1$, adding the three equations yields $0 = 3$, which shows the system of equations has no solution.

OR

We use the theorem: A system of three linear equations in three unknowns has a unique solution if and only if the determinant of its coefficient matrix does not vanish. In our example, this determinant is

$$D = \begin{vmatrix} n & 1 & 0 \\ 0 & n & 1 \\ 1 & 0 & n \end{vmatrix} = n^3 + 1.$$

The only real value of n for which D vanishes is -1. We saw above that the system has no solution in this case.

Part 2

11. (B) The given inequalities are represented geometrically as

{the set of points (x, y) such that $2\max(|x|, |y|) \leqslant a$}

$\subset \left\{$the set of points (x, y) such that $\sqrt{2(x^2 + y^2)} \leqslant a\right\}$

$\subset$ {the set of points (x, y) such that $|x| + |y| \leqslant a$}.

If the side length of the inner square in the figures is denoted by a, then the three sets in the above inclusions are the sets of points inside (or on) the inscribed square, the circle, and the circumscribed square, respectively, in Figure II.

12. (D) Let m and n denote the number of doctors and lawyers, respectively. Then

$$\frac{35m + 50n}{m + n} = 40,$$

so $40(m + n) = 35m + 50n$, $\quad 5m = 10n$, and $\dfrac{m}{n} = 2$.

13. (D) The square of the given fraction is

$$\frac{4(2 + 2\sqrt{12} + 6)}{9(2 + \sqrt{3})} = \frac{4(8 + 4\sqrt{3})}{9(2 + \sqrt{3})} = \frac{16(2 + \sqrt{3})}{9(2 + \sqrt{3})} = \frac{16}{9}.$$

Hence the fraction is equal to $\sqrt{\dfrac{16}{9}} = \dfrac{4}{3}$.

OR

Multiplying top and bottom of the given fraction by $\sqrt{2}$, we get

$$\frac{2(\sqrt{2} + \sqrt{6})}{3\sqrt{2 + \sqrt{3}}} = \frac{2(2 + \sqrt{12})}{3\sqrt{4 + 2\sqrt{3}}} = \frac{4(1 + \sqrt{3})}{3\sqrt{(1 + \sqrt{3})^2}} = \frac{4}{3}.$$

14. (C) Let x, y and z denote the number of tankfuls of water delivered by valves A, B and C, respectively, in one hour. Then

$$x + y + z = 1; \qquad x + z = \frac{1}{1.5}; \qquad y + z = \frac{1}{2}.$$

Subtracting the sum of the last two equations from twice the first yields $x + y = \frac{5}{6}$, so that $\frac{6}{5}(x + y) = 1$ tankful will be delivered by valves A and B in $\frac{6}{5} = 1.2$ hours.

15. (D) The center of the circle which circumscribes sector POQ is at C, the intersection of the perpendicular bisectors SC and RC. Considering $\triangle ORC$, we see that

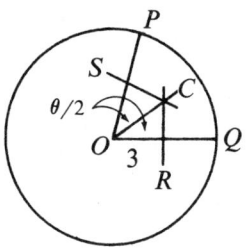

$$\sec \frac{\theta}{2} = \frac{OC}{3} \quad \text{or} \quad OC = 3 \sec \frac{\theta}{2}.$$

Note: All the answers except (D) can be eliminated by considering the limiting case $\theta = \pi/2$. For in this case (A) is 0, (B) is infinite, (C) is less than 3, and (E) is 3; all absurd.

16. (B) Let n denote the number of sides of the given convex polygon and x the number of degrees in the excepted angle. Then $180(n - 2) = 2190 + x$, so that

$$n - 2 = \frac{2190}{180} + \frac{x}{180}.$$

Since the polygon is convex, $0 < x < 180$; it follows that

$$\frac{2190}{180} < n - 2 < \frac{2190}{180} + 1,$$

i.e. $12\frac{1}{6} < n - 2 < 13\frac{1}{6}$. Since n is an integer, this forces $n - 2 = 13$, so $n = 15$. (Incidentally, $(13)(180) = 2340 = 2190 + 150$, so the excepted angle has measure $150°$.)

17. (E) Using the formula for the cosine of twice the angle $\frac{1}{2}\theta$, we have

$$\cos \theta = \cos 2\left(\frac{\theta}{2}\right) = 1 - 2 \sin^2 \frac{\theta}{2} = 1 - 2\frac{x - 1}{2x} = \frac{1}{x}.$$

[Note that since $0 < \theta < 90°$, $0 < \frac{1}{x} < 1$, so $x > 1$.] Now

$$\tan^2 \theta = \sec^2 \theta - 1 = \frac{1}{\cos^2 \theta} - 1 = x^2 - 1,$$

so

$$\tan \theta = \sqrt{x^2 - 1}.$$

18. (C) Since the factors $p - 1$ and $p + 1$ of $p^2 - 1$ are consecutive even integers, both are divisible by 2 and one of them by 4, so that their product is divisible by 8. Again $(p - 1)$, p and $(p + 1)$ are three consecutive integers, so that one of them (but not the prime p) is divisible by 3. Therefore the product $(p - 1)(p + 1) = p^2 - 1$ is always divisible by both 3 and 8, and hence by 24.

19. (D) $72_8! = 72 \times 64 \times 56 \times 48 \times 40 \times 32 \times 24 \times 16 \times 8$
$= 8^9 \times (9!);$

$18_2! = 18 \times 16 \times 14 \times 12 \times 10 \times 8 \times 6 \times 4 \times 2$
$= 2^9 \times (9!).$

The quotient $(72_8!)/(18_2!) = 8^9/2^9 = 4^9.$

20. (C) In the adjoining figure, S denotes an arbitrary point on the stream SE, and C, H and D denote the position of the cowboy, his cabin and the point 8 miles north of C, respectively. The distance $CS + SH$ equals the distance $DS + SH$, which is least when DSH is a straight line, and then

$$CSH = DSH = \sqrt{8^2 + 15^2}$$
$$= \sqrt{289} = 17 \text{ miles.}$$

Part 3

21. (B) The number of consecutive integers in a set is either odd or even. If odd, let their number be $2n + 1$ and their average be x, the middle integer. Then $(2n + 1)x = 100$ and $x = 100/(2n + 1)$, so that $2n + 1$ can only be 5 or 25. If $2n + 1 = 5$, then $x = 20$ and $n = 2$, so that the integers are $18, 19, 20, 21, 22$. If $2n + 1 = 25$, $x = 4$ and $n = 12$, which is impossible because the integers must be positive.

If the number of consecutive integers is an even number $2n$, let the average of the integers (half way between the middle pair) be denoted by x. Then $2nx = 100$, $x = 50/n$. Since x is a half integer, $2x = 100/n$ is an integer, but $x = 50/n$ is not an integer, so that n is 4, 20 or 100. For $n = 4$, $x = 12\frac{1}{2}$ and the integers are 9 through 16; $n = 20$ and 100 are impossible since the integers are positive. Hence there are exactly two sets of positive integers whose sum is 100.

22. (A) Interpreting $|x - 1| + |x + 2|$ as

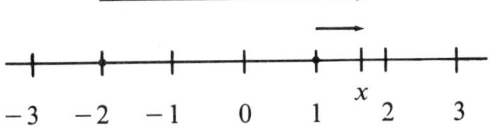

the sum of the distances from x to 1 and from x to -2, we see that this sum is 3 if $-2 < x < 1$ and $3 + 2u$ if x is a distance u from the interval from -2 to 1. Thus, the solution set is $\{x: -3 < x < 2\}$.

23. (D) Let the sides of the first card (both red) be numbered 1 and 2. Let the red and blue sides of the second card be numbered 3 and 4, respectively. On the draw any of the sides 1, 2 or 3 has equal likelihood of being face up on the table. Of these three, two undersides are red and one blue, so that the probability of a red underside is 2/3.

24. (D) Let s, c and p denote the cost in dollars of one sandwich, one cup of coffee and one piece of pie, respectively. Then $3s + 7c + p = 3.15$ and $4s + 10c + p = 4.20$. Subtracting twice the second of these equations from three times the first yields $s + c + p = 1.05$ so that $1.05 is the required cost.

25. (E) Let O denote the center of the plot and M the midpoint of the side of the walk not passing through O. If AB denotes one arc connecting opposite sides of the walk, then $OMBA$ consists of the 30° sector OAB and the 30° − 60° − 90° triangle OMB. Thus the area of the walk is

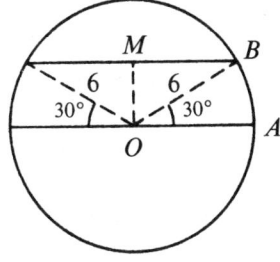

$$2\left[\frac{1}{12}\pi 6^2 + \frac{1}{2}(3)3\sqrt{3}\right] = 6\pi + 9\sqrt{3},$$

and the required area is

$$\pi 6^2 - \left(6\pi + 9\sqrt{3}\right) = 30\pi - 9\sqrt{3}.$$

26. (E) Let d and $2n$ denote the common difference and the even number of terms. Let S_o and S_e denote the sums of all n odd- and all n even-numbered terms respectively; then $S_e - S_o = nd$ since each even-numbered term exceeds its odd-numbered predecessor by d. Moreover the last term clearly exceeds the first by $(2n - 1)d$. Hence in the present case we have

$$nd = 30 - 24 = 6 \quad \text{and} \quad (2n - 1)d = 10.5.$$

Therefore $d = 2nd - 10.5 = 12 - 10.5 = 1.5$, and

$$n = 6/1.5 = 4, \quad \text{so} \quad 2n = 8.$$

Note: We note that in solving this problem we did not need to know the values of S_o and S_e, but only their difference $S_e - S_o$.

27. (A) Let s denote the distance. The total time t that car A travels is

$$t = \frac{s}{2u} + \frac{s}{2v}$$

and the average speed is

$$\frac{s}{t} = \frac{s}{\dfrac{s}{2u} + \dfrac{s}{2v}} = \frac{2uv}{u + v} = x.$$

For car B, the average speed is clearly $\frac{1}{2}(u + v) = y$. Now $0 \leqslant (u - v)^2$; $2uv \leqslant u^2 + v^2$; $4uv \leqslant (u + v)^2$; and dividing both sides of the last inequality by $2(u + v)$ yields

$$x = \frac{2uv}{u + v} \leqslant \frac{u + v}{2} = y.$$

Note 1: The solution (A) can also be obtained without any calculation, since car A clearly spends more than half the time driving at the slower of the two speeds u and v.

Note 2: Observe that car A's average speed x is the reciprocal of the average of the reciprocals of u and v; this is called the harmonic mean (H.M.) of u and v:

$$H.M.(u, v) = \frac{1}{\dfrac{1}{2}\left(\dfrac{1}{u} + \dfrac{1}{v}\right)} = \frac{2uv}{u + v}.$$

The calculations above show that the harmonic mean of two

positive numbers never exceeds their arithmetic mean. For comparisons of the harmonic, geometric and arithmetic means of two positive numbers, see An *Introduction to Inequalities* by E. Beckenbach and R. Bellman, NML vol. 3, Exercise 4 on p. 62 and its solution on p. 120.

28. (C) Let $r > 1$ denote the common ratio in the geometric progression a, b, c:

$$b = ar; \quad \log_n b = \log_n a + \log_n r$$

$$c = ar^2; \quad \log_n c = \log_n a + 2\log_n r,$$

so that $\log_n a, \log_n b, \log_n c$ is an arithmetic progression. Now it follows from the identity[†] $\log_u v = \dfrac{1}{\log_v u}$, that the reciprocals of $\log_a n, \log_b n, \log_c n$ form an arithmetic progression.

29. (A) The boys meet for the first time when the faster has covered $9/14$ and the slower $5/14$ of the track. They meet for the nth time after the faster has travelled $(9/14)n$ laps and the slower $(5/14)n$ laps. Both of these are first whole numbers of laps when $n = 14$; there are thirteen meetings, excluding the start and finish.

30. (B) For any fixed $t \geqslant 0$, $0 \leqslant T < 1$. Hence S, the interior of the circle with center $(T, 0)$ and radius T, has an area between 0 and π.

Part 4

31. (C) The integer $TTT = T(111) = T \cdot 3 \cdot 37$, with T equal to the final digit of E^2. Now 37 must divide one of ME and YE, say ME. Therefore $ME = 37$ or 74. The latter possibility is ruled out since it would imply $(ME) \cdot (YE) \geqslant 74 \cdot 14 = 1036$, while clearly $TTT \leqslant 999$. Thus $ME = 37$, $T = 9$, and $YE = \dfrac{TTT}{ME} = \dfrac{999}{37} = 27$.

[†]See footnote on p. 82.

32. (A) The volume of a pyramid is equal to one third the area of the base times the altitude. The base of the given pyramid is an equilateral triangle with sides of length 6; hence it has area $9\sqrt{3}$. Altitude h of the given pyramid may be found by applying the Pythagorean theorem to right $\triangle ABC$ in the adjoining diagram. Here B is the center of the base, so that $AB = \frac{2}{3}(3\sqrt{3}) = 2\sqrt{3}$, $h^2 = (\sqrt{15})^2 - (2\sqrt{3})^2$, and $h = \sqrt{3}$. Therefore, the volume of the pyramid is $\frac{1}{3}(9\sqrt{3})(\sqrt{3}) = 9$.

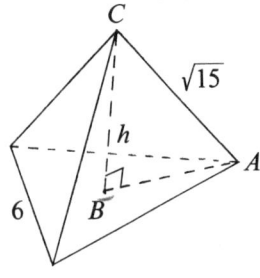

33. (C) Let x and y denote the number of ounces of water and of acid, respectively, in the original solution. After the addition of one ounce of water, there are y ounces of acid and $x + y + 1$ ounces of solution; after adding one ounce of acid, there are $y + 1$ ounces of acid and $x + y + 2$ ounces of solution. Therefore,

$$\frac{y}{x + y + 1} = \frac{1}{5} \quad \text{and} \quad \frac{y + 1}{x + y + 2} = \frac{1}{3}.$$

Solving these equations yields $x = 3$ and $y = 1$, from which it follows that the original solution contained

$$\frac{y}{x + y} = \frac{1}{3 + 1} = 25\% \text{ acid.}$$

34. (C) Let d, v and w denote the distance between the towns, the speed against the wind and the speed with the wind, respectively; then the plane's speed in still air is $\frac{1}{2}(v + w)$. We are given that

(1) $\dfrac{d}{v} = 84$ and that (2) $\dfrac{d}{w} = \dfrac{d}{\frac{1}{2}(w + v)} - 9.$

Set $x = d/w$, the required return time. Then by (2),

$$x = \frac{2}{\dfrac{w}{d} + \dfrac{v}{d}} - 9 = \frac{2}{\dfrac{1}{x} + \dfrac{1}{84}} - 9.$$

Simplifying, we obtain $x^2 - 75x + 756 = 0$. The solutions of this equation are $x = 63$ and $x = 12$.

35. (E) We shall show that I, II, and III are all true by giving geometric arguments for I and II and then showing algebraically that III is a consequence of I and II.

In the adjoining figure, chords QN and KM have length s. The five equal chords of length s in the semicircle with diameter KR subtend central angles of $180°/5 = 36°$ each. The five isosceles triangles, each with base s and opposite vertex at the center O, have base angles of measure

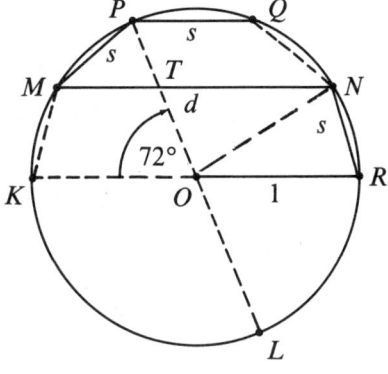

$$\tfrac{1}{2}(180° - 36°) = 72°.$$

Now rotate the entire configuration clockwise through $72°$ about O. Then chord PQ, parallel to diameter KR, goes into chord NR, parallel to diameter PL. Denote by T the intersection of MN and PL. In parallelogram $ORNT$, $TN = OR = 1$, and $TO = NR = s$. We saw that $\angle MPO = 72°$; now $\angle MTP = 72°$ also, because $MT \| KO$. So $\triangle PMT$ is isosceles with $MT = MP = s$. Therefore $d = MT + TN = s + 1$, so

I $\qquad\qquad\qquad d - s = 1.$

The segments PT, TL and MT, TN of intersecting chords MN and PL satisfy $(PT)(TL) = (MT)(TN)$. Since $PT = OP - OT = 1 - s$, and $TL = OL + OT = 1 + s$, this equation takes the form

$$(1 - s)(1 + s) = s \cdot 1 \quad \text{or} \quad 1 - s^2 = s.$$

Multiplying equation (I) by s, we get $ds - s^2 = s$, so

II $\qquad\qquad\qquad ds = s^2 + s = 1.$

The equation $s^2 + s = 1$ is equivalent to $s^2 + s - 1 = 0$, whose positive root is $s = \tfrac{1}{2}(\sqrt{5} - 1)$. Therefore

$$d = s + 1 = \frac{\sqrt{5}}{2} + \frac{1}{2},$$

and

III $\qquad d^2 - s^2 = (d + s)(d - s) = d + s = \sqrt{5}.$

OR

Each chord of length s subtends an angle $36°$ at the center O, and MN of length d subtends an angle $3 \cdot 36° = 108°$. Therefore

$$s = 2\sin 18° \quad \text{and} \quad d = 2\sin 54°.$$

It follows from this and trigonometric identities that

$$d = 2\sin 54° = 2\cos 36° = 2(1 - 2\sin^2 18°) = 2 - s^2$$

and

$$s = 2\sin 18° = 2\cos 72° = 2(\cos^2 36° - 1) = d^2 - 2.$$

Adding $d = 2 - s^2$ and $s = d^2 - 2$, we obtain $d + s = d^2 - s^2 = (d + s)(d - s)$, so that I. $d - s = 1$. Substituting $s + 1$ for d in $d = 2 - s^2$, we find that $s^2 + s - 1 = 0$, which has one positive root, $s = \frac{1}{2}(\sqrt{5} - 1)$. Then

$$d = s + 1 = \frac{\sqrt{5} + 1}{2}, \qquad ds = 1, \qquad d + s = \sqrt{5}$$

and

$$d^2 - s^2 = \sqrt{5}.$$

1974 Solutions

1. (D) Multiplying both sides of the given equation by the least common denominator $2xy$ yields $4y + 6x = xy$ or, equivalently, $4y = xy - 6x$. Factoring x from the right side of the last equation gives $4y = x(y - 6)$. Since $y \neq 6$ we can divide both sides of this equation by $y - 6$ to obtain (D).

2. (B) Since x_1 and x_2 are the two roots of the quadratic equation $3x^2 - hx - b = 0$, the sum of the roots is $\dfrac{h}{3}$.

3. (A) The coefficient of x^7 in $(1 + 2x - x^2)^4$ is the coefficient of the sum of four identical terms $2x(-x^2)^3$, which sum is $-8x^7$.

4. (D) By the remainder theorem $x^{51} + 51$ divided by $x + 1$ leaves a remainder of $(-1)^{51} + 51 = 50$. This can also be seen quite easily by long division.

5. (B) $\angle EBC = \angle ADC$ since both angles are supplements of $\angle ABC$. Note the fact that $\angle BAD = 92°$ is not needed in the solution of the problem.

6. (D) $x * y = \dfrac{xy}{x + y} = \dfrac{yx}{y + x} = y * x$

   and

   $$\left(x * y\right) * z = \dfrac{xy}{x + y} * z = \dfrac{\dfrac{xyz}{x + y}}{\dfrac{xy}{x + y} + z} = \dfrac{xyz}{xy + xz + yz}.$$

   Similarly $x * (y * z) = \dfrac{xyz}{xy + xz + yz}$, so "$*$" is both commutative and associative.

   Note: The result can also be derived from the identity

   $$\dfrac{1}{x * y} = \dfrac{1}{x} + \dfrac{1}{y}.$$

   Commutativity and associativity of $*$ now follows from the fact that they hold for $+$.

7. (D) If x is the original population, then
$$x + 1{,}200 - 0.11(x + 1{,}200) = x - 32.$$
Solving for x gives (D).

8. (A) Since 3^{11} and 5^{13} are both odd, their sum is even.

9. (B) All multiples of 8, including 1,000, fall in the second column.

10. (B) Putting the quadratic in its standard form:
$$(2k - 1)x^2 - 8x + 6 = 0,$$
we see that the discriminant D is $64 - 4(2k - 1)6 = 88 - 48k = 8(11 - 6k)$. A quadratic equation has no real roots if and only if its discriminant is negative. D is negative if $11 - 6k < 0$, that is, when $k > 11/6$; the smallest *integral* value of k for which the equation has no real roots is 2.

11. (A) Since (a, b) and (c, d) are on the same line, $y = mx + k$, they satisfy the same equation. Therefore,
$$b = ma + k \qquad d = mc + k.$$
Now the distance between (a, b) and (c, d) is
$$\sqrt{(a - c)^2 + (b - d)^2}.$$
From the first two equations we obtain $(b - d) = m(a - c)$, so that
$$\sqrt{(a - c)^2 + (b - d)^2} = \sqrt{(a - c)^2 + m^2(a - c)^2}$$
$$= |a - c|\sqrt{1 + m^2}.$$
Note we are using the fact that $\sqrt{x^2} = |x|$ for all real x.

OR

Let θ be the angle between the line and the x-axis; then
$$(\tan \theta)^2 = m^2 = \frac{(b - d)^2}{(a - c)^2},$$
so
$$(b - d)^2 = m^2(a - c)^2,$$
and the square of the desired distance is
$$(a - c)^2 + (b - d)^2 = (a - c)^2 + m^2(a - c)^2$$
$$= (a - c)^2(1 + m^2).$$

12. (B) Since $g(x) = 1/2$ is satisfied by $x = \sqrt{1/2}$,

$$f(1/2) = f\left(g\left(\sqrt{1/2}\right)\right) = 1.$$

OR

Since $g(x) = 1 - x^2$ for $x \neq 0$, we have $x^2 = 1 - g(x)$ for $x \neq 0$; so

$$\frac{1 - x^2}{x^2} = \frac{g(x)}{1 - g(x)} = f(g(x)) \quad \text{for } x \neq 0,$$

and

$$f\left(\tfrac{1}{2}\right) = \frac{\tfrac{1}{2}}{1 - \tfrac{1}{2}} = 1.$$

13. (D) Statement (D) is the contrapositive of the given one and the only one of the statements (A) through (E) equivalent to the given statement.

14. (A) Since $x^2 > 0$ for all $x \neq 0$, $x^2 > 0 > x$ is true if $x < 0$. Counterexamples to the other statements are easy to construct.

15. (B) By definition $|a| = \left\{ \begin{array}{l} a \text{ when } a \geqslant 0 \\ -a \text{ when } a < 0 \end{array} \right.$. If $x < -2$, then $1 + x < 0$ and $|1 + x| = -(1 + x)$ and $|1 - |1 + x||$ $= |1 + 1 + x| = |2 + x|$. Again if $x < -2$, then $2 + x < 0$ and $|2 + x| = -2 - x$.

16. (A) In the adjoining figure, $\triangle ABC$ is a right isosceles triangle, with $\angle BAC = 90°$ and $AB = AC$, inscribed in a circle with center O and radius R. The line segment AO has length R and bisects line segment BC and $\angle BAC$. A circle with center O' lying on AO and radius r is inscribed in $\triangle ABC$. The sides AB 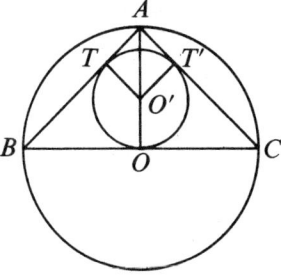 and AC are tangent to the inscribed circle with points of tangency T and T', respectively. Since $\triangle ATO'$ has angles $45° - 45° - 90°$ and $O'T = r$, we have $AT = r$ and $O'A = R - r = r\sqrt{2}$. Hence $R = r + r\sqrt{2}$ and $R/r = 1 + \sqrt{2}$.

17. (C) Since $i^2 = -1$, $(1 + i)^2 = 2i$ and $(1 - i)^2 = -2i$. Writing

$$(1 + i)^{20} - (1 - i)^{20} = \left((1 + i)^2\right)^{10} - \left((1 - i)^2\right)^{10},$$

we have

$$(1 + i)^{20} - (1 - i)^{20} = (2i)^{10} - (-2i)^{10} = 0.$$

OR

Observe that $i(1 - i) = i - i^2 = 1 + i$, and $i^4 = (-1)^2 = 1$. Therefore

$$(1 + i)^4 = i^4(1 - i)^4 = (1 - i)^4,$$

and

$$(1 + i)^{4n} - (1 - i)^{4n} = 0 \quad \text{for} \quad n = 1, 2, 3, \ldots .$$

OR

$$(1 + i)^{20} = \sqrt{2}^{\,20}\left[\cos 900° + i \sin 900°\right] = -1024,$$

$$(1 - i)^{20} = \sqrt{2}^{\,20}\left[\cos(-900°) + i \sin(-900°)\right] = -1024,$$

and the difference is 0.

18. (D) By hypothesis we have

$$3 = 8^p = 2^{3p}, \qquad 5 = 3^q; \quad \text{so } 5 = \left(2^{3p}\right)^q = 2^{3pq}.$$

We are asked to find $\log_{10} 5 = x$, i.e. x such that $10^x = 5$. Since $5 = 2^{3pq}$, we have $10^x = 2^{3pq}$; but also $10^x = 2^x \cdot 5^x$ $= 2^x \cdot 2^{3pqx} = 2^{x(1 + 3pq)}$. It follows that

$$x(1 + 3pq) = 3pq \quad \text{and} \quad x = \frac{3pq}{1 + 3pq}.$$

OR

We recall the change of base formula[†]

$$(*) \quad \log_a b = \frac{\log_c b}{\log_c a} \quad \text{and its corollary} \quad \log_a b = \frac{1}{\log_b a}$$

[†] To verify the change of base formula $(*)$ for positive numbers $a, b, c, a \neq 1, c \neq 1$ we set

$$\log_a b = U, \quad \text{so} \quad a^U = b,$$
$$\log_c b = V, \quad \text{so} \quad c^V = b,$$
$$\log_c a = W, \quad \text{so} \quad c^W = a.$$

Thus $c^V = b = a^U = (c^W)^U = c^{UW}$ and $UW = V$, so $U = V/W$ as claimed. Taking $b = c$, we also obtain the important corollary

$$\log_a b = \frac{1}{\log_b a}.$$

and write

$$p = \log_8 3 = \frac{1}{\log_3 8} = \frac{1}{\log_3 2^3} = \frac{1}{3\log_3 2},$$

so

$$\log_3 2 = \frac{1}{3p},$$

and

$$\log_{10} 5 = \frac{\log_3 5}{\log_3 10} = \frac{q}{\log_3 2 + \log_3 5} = \frac{q}{\frac{1}{3p} + q} = \frac{3pq}{1 + 3pq}.$$

OR

Since we are asked for an answer in base 10, we convert all the given information to that base, using (∗). From

$$p = \log_8 3 = \frac{\log_{10} 3}{\log_{10} 8}, \qquad q = \log_3 5 = \frac{\log_{10} 5}{\log_{10} 3}$$

one obtains

$$\log_{10} 3 = p\log_{10} 8, \qquad \log_{10} 5 = q\log_{10} 3,$$

thus

$$\log_{10} 5 = pq\log_{10} 8 = pq\log_{10}\left(\frac{10}{5}\right)^3,$$

$$= 3pq(1 - \log_{10} 5).$$

Solving this for $\log_{10} 5$ gives choice (D).

Note: Conversion of all logarithms to a common base provides a systematic (if not always shortest) approach to problems like 18, and this approach will be used again in later problems, with a reference back to (∗). In the last solution to this problem, we converted to base 10 only because the answer was requested in that base. The traditional reason for using base 10 – ease of doing numerical computations – has been made largely obsolete by computers.

19. (A) Let $DM = NB = x$; then $AM = AN = 1 - x$. Now

$$\text{area } \triangle CMN = \text{area } \square ABCD - \text{area } \triangle ANM$$
$$- \text{area } \triangle NBC - \text{area } \triangle CDM$$
$$= 1 - \frac{1}{2}(1 - x)^2 - \frac{x}{2} - \frac{x}{2} = \frac{1}{2}(1 - x^2).$$

Denote the length of each side of the equilateral triangle CMN by y; using the Pythagorean theorem we see that

$$x^2 + 1^2 = y^2 \quad \text{and} \quad (1 - x)^2 + (1 - x)^2 = y^2.$$

Substituting the first equation into the second we get

$$2(1 - x)^2 = x^2 + 1 \quad \text{or} \quad x^2 - 4x + 1 = 0.$$

The roots of this equation are $2 - \sqrt{3}$ and $2 + \sqrt{3}$. Since $2 + \sqrt{3} > 1$ we must choose $x = 2 - \sqrt{3}$ and obtain area $\triangle CMN = 2\sqrt{3} - 3$.

20. (D) By rationalizing the denominator of each fraction, we see

$$T = \left(3 + \sqrt{8}\right) - \left(\sqrt{8} + \sqrt{7}\right) + \left(\sqrt{7} + \sqrt{6}\right)$$
$$- \left(\sqrt{6} + \sqrt{5}\right) + \left(\sqrt{5} + 2\right)$$
$$= 3 + 2 = 5.$$

21. (B) The sum of the first five terms of the geometric series with initial term a and common ratio r is

$$S_5 = a + ar + ar^2 + ar^3 + ar^4 = \frac{a(1 - r^5)}{1 - r}.$$

By hypothesis $ar^4 - ar^3 = 576$ and $ar - a = 9$. Dividing the first equation by the last yields

$$\frac{r^4 - r^3}{r - 1} = \frac{r^3(r - 1)}{r - 1} = 64,$$

so $r^3 = 64$, and $r = 4$. Since $ar - a = 9$ and $r = 4$, we have $a = 3$ and therefore

$$S_5 = \frac{3(1 - 4^5)}{-3} = 1023.$$

22. (E) We recall that $a \sin \theta + b \cos \theta$, if $a^2 + b^2 \neq 0$, can be expressed in the form

$$\sqrt{a^2 + b^2} \sin(\theta + \varphi),$$

where
$$\cos \varphi = \frac{a}{\sqrt{a^2 + b^2}}, \qquad \sin \varphi = \frac{b}{\sqrt{a^2 + b^2}},$$
and that the minimum of $\sin \alpha$ is -1 and occurs when $\alpha = 270° + (360m)°$, $m = 0, \pm 1, \pm 2, \ldots$.

Applying this principle to the given function where
$$\frac{a}{\sqrt{a^2 + b^2}} = \frac{1}{\sqrt{1 + 3}} = \frac{1}{2}, \qquad \frac{b}{\sqrt{a^2 + b^2}} = -\frac{\sqrt{3}}{2},$$
we find $\varphi = -60°$ and write
$$\sin\frac{A}{2} - \sqrt{3}\cos\frac{A}{2} = 2\left[\frac{1}{2}\sin\frac{A}{2} - \frac{\sqrt{3}}{2}\cos\frac{A}{2}\right]$$
$$= 2\sin\left(\frac{A}{2} - 60°\right).$$

This expression is minimized when $\frac{1}{2}A - 60° = 270° + (360m)°$, that is when $A = 660° + (720m)°$, $m = 0$, $\pm 1, \pm 2, \ldots$. None of (A) through (D) are angles of this form.

23. (B) Since $TP = T''P$, $OT = OT'' = r$, and $\angle PT''O = \angle PTO = 90°$, we have $\triangle OTP \cong \triangle OT''P$. Similarly $\triangle OT'Q \cong \triangle OT''Q$. Letting $x = \angle TOP = \angle POT''$ and $y = \angle T''OQ = \angle QOT'$ we obtain $2x + 2y = 180°$. But this implies that $\angle POQ = x + y = 90°$. Therefore $\triangle POQ$ is a right triangle with altitude OT''. Since the altitude drawn to the hypotenuse of a right triangle is the mean proportion of the segments it cuts, we have
$$\frac{4}{r} = \frac{r}{9} \quad \text{or} \quad r = 6.$$

24. (A) Let A be the event of rolling at least a five; then the probability of A is $\frac{2}{6} = \frac{1}{3}$. In six rolls of a die the probability of event A happening six times is $\left(\frac{1}{3}\right)^6$. The probability of getting exactly five successes and one failure of A in a specific order is $\left(\frac{1}{3}\right)^5 \cdot \frac{2}{3}$. Since there are six ways to do this, the probability of getting five successes and one failure of A in any order is
$$6\left(\frac{1}{3}\right)^5 \cdot \frac{2}{3} = \frac{12}{729}.$$
The probability of getting all successes or five successes and one failure in any order is thus
$$\frac{12}{729} + \left(\frac{1}{3}\right)^6 = \frac{13}{729}.$$

25. (C) Since $DM = AM, \angle QMA = \angle DMC$ and $\angle CDM = \angle QAM$, we have $\triangle QAM \cong \triangle MCD$. Similarly $\triangle BPN \cong \triangle DNC$. Now,

$$\text{area } \triangle QPO = \text{area } \square ABCD + \text{area } \triangle DOC$$

and

$$\text{area } \triangle DOC = \frac{1}{4}\left(\frac{1}{2}\text{area } \square ABCD\right) = \frac{k}{8},$$

so that

$$\text{area } \triangle QPO = k + \frac{k}{8} = \frac{9k}{8}.$$

26. (C) Writing 30 as a product of prime factors, $30 = 2 \cdot 3 \cdot 5$, we obtain

$$(30)^4 = 2^4 \cdot 3^4 \cdot 5^4.$$

The divisors of $(30)^4$ are exactly the numbers of the form $2^i \cdot 3^j \cdot 5^k$, where i, j, k are non-negative integers between zero and four inclusively, so there are $(5)^3 = 125$ distinct divisors of $(30)^4$; excluding 1 and $(30)^4$ there are 123 divisors.

27. (A) Consider $|f(x) + 4| = |3x + 2 + 4| = 3|x + 2|$. Now whenever $|x + 2| < a/3$, then $|f(x) + 4| < a$. Consequently whenever $|x + 2| < b$ and $b \leq a/3$, we have $|f(x) + 4| < a$.

28. (D) Using the formula for the sum of a geometric series, we have

$$0 \leq x \leq \sum_{n=1}^{25} \frac{2}{3^n} < \sum_{n=1}^{\infty} \frac{2}{3^n} = \frac{\frac{2}{3}}{1 - \frac{1}{3}} = 1.$$

If $a_1 = 0$, then

$$0 \leq x < \sum_{n=2}^{\infty} \frac{2}{3^n} = \frac{\frac{2}{9}}{1 - \frac{1}{3}} = \frac{1}{3}.$$

If $a_1 = 2$, then

$$\frac{2}{3} + 0 + 0 + \cdots \leq x < 1.$$

So either $0 \leq x < \frac{1}{3}$ or $\frac{2}{3} \leq x < 1$.

OR

If the number x is represented in base 3, then it takes the form $.a_1a_2a_3 \cdots a_{25}$ where each digit is either 0 or 2. Thus we see that either

$$x < (0.100\ldots)_3 = \frac{1}{3} \quad \text{or} \quad x \geqslant (0.200\ldots)_3 = \frac{2}{3}.$$

29. (B) For each $p = 1,\ldots, 10$,

$$S_p = p + p + (2p - 1) + p + 2(2p - 1) + \cdots + p + 39(2p - 1)$$

$$= 40p + \frac{(40)(39)}{2}(2p - 1)$$

$$= (40 + 40 \cdot 39)p - 20 \cdot 39$$

$$= (40)^2 p - 20 \cdot 39.$$

Therefore,

$$\sum_{p=1}^{10} S_p = (40)^2 \sum_{p=1}^{10} p - (10)(20)(39)$$

$$= (40)^2 \frac{(10)(11)}{2} - (10)(20)(39)$$

$$= 80,200.$$

30. (A) Consider line segment AB cut by a point D with $AD = x$, $DB = y$, $y < x$ and

$$\frac{y}{x} = \frac{x}{x + y}.$$

Since $y/x = R$ we can choose $x = 1$ and therefore $y = R$; thus $R = \dfrac{1}{1 + R}$ and $R^2 + R - 1 = 0$. We can therefore write $R^{-1} = R + 1$ so that

$$R^2 + R^{-1} = R^2 + R + 1 = (R^2 + R - 1) + 2 = 2,$$

and

$$R^{[R^{(R^2+R^{-1})}+R^{-1}]} + R^{-1} = R^{[R^2+R^{-1}]} + R^{-1}$$

$$= R^2 + R^{-1}$$

$$= 2.$$

1975 Solutions

1. (B) $\dfrac{1}{2 - \dfrac{1}{2 - \dfrac{1}{2 - \dfrac{1}{2}}}} = \dfrac{1}{2 - \dfrac{1}{2 - \dfrac{2}{3}}} = \dfrac{1}{2 - \dfrac{3}{4}} = \dfrac{4}{5}.$

2. (D) The equations have a solution unless their graphs are parallel lines. This will be the case only if their slopes are equal, i.e. if $m = 2m - 1$ or $m = 1$. (The lines are not coincident since they have distinct y-intercepts for all values of m.)

3. (A) None of the inequalities are satisfied if a, b, c, x, y, z are chosen to be $1, 1, -1, 0, 0, -10$, respectively.

4. (A) In the adjoining figure, if s is the length of a side of the first square, then $s/\sqrt{2}$ is the length of a side of the second. Thus the ratio of the areas is $s^2/(s/\sqrt{2})^2 = 2$.

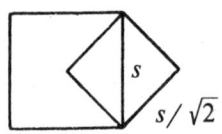

5. (B) $(x + y)^9 = x^9 + 9x^8y + 36x^7y^2 + \cdots + y^9;$

   $9p^8q = 36p^7q^2$ and $p + q = 1$. Dividing the second equation by $9p^7q$, we obtain $p = 4q$, and substituting $1 - p$ for q we obtain $p = 4(1 - p)$, so $p = 4/5$.

6. (E) Grouping the terms of the difference as
   $$(2 - 1) + (4 - 3) + \cdots + (160 - 159),$$
   one obtains a sum of 80 terms, each equal to 1.

7. (E) $\dfrac{|x - |x||}{x}$ is -2 if x is negative and 0 if x is positive.

8. (D) II and IV are the only negations of the given statement.

9. (C) Let d denote the common difference of the progression $a_1 + b_1$, $a_2 + b_2$, ... Then

$$99d = (a_{100} + b_{100}) - (a_1 + b_1) = 0.$$

Thus $d = 0$, and $100(a_1 + b_1) = 10{,}000$ is the desired sum.

10. (A) Let k be any positive integer. Then

$$(10^k + 1)^2 = 10^{2k} + 2 \cdot 10^k + 1.$$

The sum of the digits is therefore $1 + 2 + 1 = 4$.

11. (E) Suppose P is the given point, O is the center of circle K, and M is the midpoint of a chord AB passing through P. Since $\angle OMP$ is $90°$, M lies on a circle C having OP for its diameter. Conversely, if M is any point on the circle C, then the chord of circle K passing through P and M (the chord of K tangent to C at P if $M = P$) is perpendicular to OM. Hence M is the midpoint of this chord and therefore belongs to the locus.

12. (B) If $a \neq b$, $a^3 - b^3 = 19x^3$ and $a - b = x$, then

$$a^3 - b^3 = (a - b)(a^2 + ab + b^2) = x(a^2 + ab + b^2)$$
$$= 19x^3$$

Dividing the last equality above by x and substituting $b = a - x$, we obtain

$$18x^2 + 3ax - 3a^2 = 0,$$
$$-3(a - 3x)(a + 2x) = 0.$$

So $a = 3x$ or $a = -2x$.

13. (D) For $x < 0$, the polynomial $x^6 - 3x^5 - 6x^3 - x + 8$ is positive, since then all terms are positive; so it has no negative zeros. At $x = 1$, the polynomial is negative and hence has at least one positive zero (between 0 and 1).

14. (E) Let W, H, I and S denote *whatsis*, *whosis*, *is* and *so*, respectively. Then $H = I$ and $IS = 2S$ imply $W = S$, or equivalently, since $S > 0$, $H = I = 2$ implies $W = S$. Now if $H = S$, $2S = S^2$ and $I = 2$, or equivalently $H = I = S = 2$, then $W = S$, so that $HW = 4 = S + S$.

15. (A) The first eight terms of the sequence are 1, 3, 2, -1, -3, -2, 1, 3. Since the seventh and eighth terms are the same as the first and second, the ninth term will be the same as the third, etc.; i.e., the sequence repeats every six terms. Moreover, the sum of each six term period is 0. Hence the sum of the first 96 terms is zero, and the sum of the first one hundred terms is the sum of the last four terms $1 + 3 + 2 - 1 = 5$.

16. (C) Denote the first term of the series by a and the common ratio by $1/n$; then the sum of the series is

$$\frac{a}{1 - (1/n)} = 3, \quad \text{and} \quad a = 3 - (3/n).$$

Since a and n are positive integers, $0 < 1/n < 1$, $n = 3$ and $a = 2$. The sum of the first two terms is $2 + 2(1/3) = 8/3$.

Note: The desired sum clearly lies between 1 and 3, so (A), (B) and (E) are impossible. If (D) were the answer, the first two terms would both be 1, also impossible. This leaves only (C).

17. (D) Since the commuter makes two trips each work day, the total number of trips is $2x$; thus $2x = 9 + (8 + 15) = 32$, and $x = 16$.

Note: The given information was deliberately redundant. "If he comes home on the train, he took the bus in the morning" is logically equivalent to "If he takes the train in the morning, he comes home by bus in the afternoon."

18. (D) There are 900 three digit numbers, and three of them (128, 256 and 512) have logarithms to base two which are integral. So $3/900 = 1/300$ is the desired probability.

19. (D) For any fixed positive value of x distinct from one, let $a = \log_3 x$, $b = \log_x 5$ and $c = \log_3 5$. Then $x = 3^a$, $5 = x^b$ and $5 = 3^c$. These last equalities imply $3^{ab} = 3^c$ or $ab = c$. Note that $\log_x 5$ is not defined for $x = 1$.

OR

Converting all these logs to some fixed but arbitrary base d (see footnote on p. 82), we obtain

$$(\log_3 x)(\log_x 5) = \frac{\log_d x}{\log_d 3} \cdot \frac{\log_d 5}{\log_d x} = \frac{\log_d 5}{\log_d 3} = \log_3 5,$$

for all $x \neq 1$.

20. (B) In the adjoining figure let h be the length of altitude AN drawn to BC, let $x = BM$ and let $y = NM$. Then

$$h^2 + (x + y)^2 = 64,$$
$$h^2 + y^2 \qquad = 9,$$
$$h^2 + (x - y)^2 = 16.$$

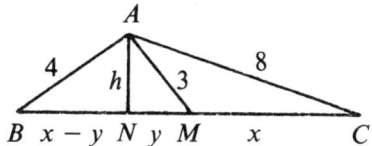

Subtracting twice the second equation from the sum of the first and third equations yields $2x^2 = 62$. Thus $x = \sqrt{31}$ and $BC = 2\sqrt{31}$.

OR

Recall that the sum of the squares of the sides of a parallelogram is equal to the sum of the squares of its diagonals. Applying this to the parallelogram having AB and AC as adjacent sides yields $2(4^2 + 8^2) = 6^2 + (2x)^2$, $x = \sqrt{31}$.

21. (D) Letting $a = 0$ in the equation $f(a)f(b) = f(a + b)$ (called a *functional equation*) yields $f(0)f(b) = f(b)$, or $f(0) = 1$; letting $b = -a$ in the functional equation yields $f(a)f(-a) = f(0)$, or $f(-a) = 1/f(a)$; and

$$f(a)f(a)f(a) = f(a)f(2a) = f(3a), \quad \text{or} \quad f(a) = \sqrt[3]{f(3a)}.$$

The function $f(x) \equiv 1$ satisfies the functional equation, but does not satisfy condition IV.

22. (E) Since the product of the positive integral roots is the prime integer q, q must be positive and the roots must be 1 and q. Since $p = 1 + q$ is also prime, $q = 2$ and $p = 3$. Hence all four statements are true.

23. (C) In the adjoining figure diagonals
    AC and DB are drawn. Since O
    is the intersection of the medians
    of $\triangle ABC$, the altitude of $\triangle AOB$
    from O is $\frac{1}{3}$ the altitude of
    $\triangle ABC$ from C; i.e. $\frac{1}{3}$ the side
    length, s, of the square. Hence

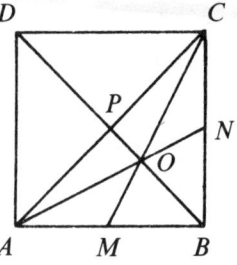

$$\text{area } \triangle AOB = \tfrac{1}{3}\left(\text{area } \triangle ABC\right)$$
$$= \tfrac{1}{3}\left(\tfrac{1}{2}s^2\right) = \tfrac{1}{6}s^2.$$

Similarly, area $\triangle COB = \frac{1}{6}s^2$. The area of $AOCD$ is obtained by subtracting the areas of triangles AOB and COB from that of the square, so area $AOCD = s^2 - \frac{1}{3}s^2 = \frac{2}{3}s^2$.

OR

Introduce coordinates with respect to which AB is the unit interval on the positive x-axis and AD is the unit interval on the positive y-axis. Now

$$\text{area } AOCD = \text{area } \triangle ACD + \text{area } \triangle AOC$$
$$= \frac{1}{2} + \frac{1}{2}\begin{vmatrix} \frac{2}{3} & \frac{1}{3} \\ 1 & 1 \end{vmatrix} = \frac{2}{3}.$$

The rows of the determinant are the coordinates of O and C. Those of O are $(\frac{2}{3}, \frac{1}{3})$ because O is the intersection of medians of $\triangle ABC$.

Note: Since the area of $\triangle ABN$ is $\frac{1}{4}$ of the area of the square, it is clear that the desired ratio r satisfies $\frac{1}{2} < r < \frac{3}{4}$. Only (C) fulfills this condition.

24. (E) If $0° < \theta < 45°$, then (see Figure 1) an application of a theorem on exterior angles of triangles to $\triangle EAC$ yields $2\theta = \angle EAC + \theta$. Therefore $\angle EAC = \theta$ and $\triangle EAC$ is isosceles. Hence $EC = AE = AD$.

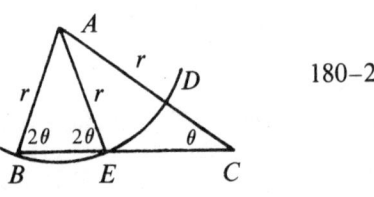

Figure 1

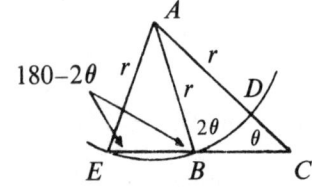

Figure 2

If $\theta = 45°$, then $\triangle ABC$ is a $45°$–$45°$–$90°$ triangle and $E = B$. Then $EC = BC = AB = AD$.

If $45° < \theta < 60°$, then (see Figure 2)

$$\measuredangle EAC = 180° - \measuredangle AEC - \measuredangle C$$
$$= 180° - (180° - 2\theta) - \theta$$
$$= \theta.$$

Thus $\triangle EAC$ is isosceles and $EC = EA = AD$.

25. (B) If the son is the worst player, the daughter must be his twin. The best player must then be the brother. This is consistent with the given information, since the brother and the son could be the same age. The assumption that any of the other players is worst leads to a contradiction:

If the woman is the worst player, her brother must be her twin and her daughter must be the best player. But the woman and her daughter cannot be the same age.

If the brother is the worst player, the woman must be his twin. The best player is then the son. But the woman and her son cannot be the same age, and hence the woman's twin, her brother, cannot be the same age as the son.

If the daughter is the worst player, the son must be the daughter's twin. The best player must then be the woman. But the woman and her daughter cannot be the same age.

26. (C) In the adjoining figure

$$BD/CD = AB/AC,$$

since an angle bisector of a triangle divides the opposite side into segments which are proportional to the two adjac-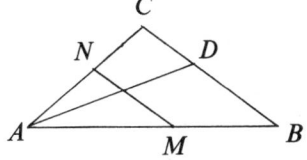
ent sides. Since $CN = CD$ and $BM = BD$, we have $BM/CN = AB/AC$, which implies that MN is parallel to BC. Since only one choice is correct, it must therefore be (C). Actually, it is easy to verify that (A), (B), (D) and (E) are false if $\measuredangle A = 90° - \delta$, $\measuredangle B = 60°$ and $\measuredangle C = 30° + \delta$, where δ is any sufficiently small positive angle.

27. (E) If p, q, r are roots, then the polynomial can be factored as follows:

$$x^3 - x^2 + x - 2 = (x - p)(x - q)(x - r)$$
$$= x^3 - (p + q + r)x^2$$
$$+ (pq + pr + qr)x - pqr.$$

Equating coefficients of like powers of x, we find

$$p + q + r = 1, \quad pq + pr + qr = 1, \quad pqr = 2.$$

In looking for the sum of the cubes of the roots of a cubic equation, let us use the fact that each root satisfies the equation:

$$p^3 - p^2 + p - 2 = 0$$
$$q^3 - q^2 + q - 2 = 0$$
$$r^3 - r^2 + r - 2 = 0.$$

Adding these, we obtain

$$(*) \quad p^3 + q^3 + r^3 - (p^2 + q^2 + r^2) + (p + q + r) - 6 = 0.$$

We saw that $p + q + r = 1$ and shall determine the sum of the squares of the roots by squaring this relation:

$$(p + q + r)^2 = p^2 + q^2 + r^2 + 2(pq + pr + qr) = 1$$
$$p^2 + q^2 + r^2 + 2(1) \qquad\qquad = 1,$$
$$p^2 + q^2 + r^2 \qquad\qquad\quad = -1.$$

Substituting this into $(*)$, we obtain

$$p^3 + q^3 + r^3 = -1 - 1 + 6 = 4.$$

28. (A) Construct line CP parallel to EF and intersecting AB at P. Then

$$\frac{AP}{AC} = \frac{AF}{AE},$$

that is,

$$\frac{AP}{16} = \frac{AF}{2AF},$$

so

$$AP = 8.$$

Let $a, x, y, \alpha, \beta, \delta$ and θ be as shown in the adjoining diagram. The desired ratio EG/GF is the same as y/x which we now determine. By the law of sines,

$$\frac{a}{\sin \alpha} = \frac{12}{\sin \theta} \, ; \qquad \frac{a}{\sin \beta} = \frac{16}{\sin(180° - \theta)} = \frac{16}{\sin \theta} \, .$$

Hence

$$\frac{\sin \beta}{\sin \alpha} = \frac{3}{4} \, .$$

Moreover

$$\frac{x}{\sin \alpha} = \frac{8}{\sin \delta} \, ; \qquad \frac{y}{\sin \beta} = \frac{16}{\sin(180° - \delta)} = \frac{16}{\sin \delta} \, .$$

Hence

$$\frac{y}{x} = 2\frac{\sin \beta}{\sin \alpha} = \frac{3}{2} \, .$$

OR

Join GB and GC. Triangle ABC is subdivided into six smaller triangles whose areas are denoted by a, b, c, d, e, f, as indicated in the diagram. Triangles AEG and AFG have the common vertex A, so their areas are in the ratio EG to GF. Thus

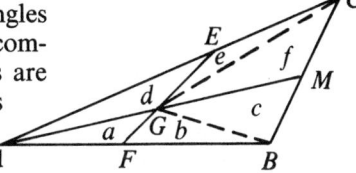

$$\frac{EG}{GF} = \frac{d}{a} \, ,$$

and this we now calculate.

Triangles ACM and ABM have equal areas, so

$$d + e + f = a + b + c.$$

Similarly $f = c$, and hence $d + e = a + b$ by subtraction. Let x be the length AF, so that $AE = 2x$, $FB = 12 - x$, $EC = 16 - 2x$. Then

$$\frac{b}{a} = \frac{FB}{FA} = \frac{12 - x}{x} \quad \text{and} \quad \frac{e}{d} = \frac{EC}{EA} = \frac{16 - 2x}{2x} \, .$$

Hence

$$a + b = a + \frac{12 - x}{x}a = a\frac{12}{x}$$

and

$$d + e = d + \frac{16 - 2x}{2x}d = d\frac{16}{2x} \, .$$

Thus $a + b = d + e$ becomes

$$a\frac{12}{x} = d\frac{16}{2x}, \quad \text{or} \quad 3a = 2d, \quad \text{so} \quad \frac{d}{a} = \frac{3}{2}.$$

OR

Extend BC and FE until they intersect in a point H; see the adjoining figure. The collinear points A, G, M lie on sides (or their extensions) BF, FH, HB of $\triangle FBH$ respectively. They also lie on extensions of sides CE, EH, HC of $\triangle ECH$. We may therefore apply Menelaus's theorem[†] and find

$$\frac{HG}{FG} \cdot \frac{FA}{BA} \cdot \frac{BM}{HM} = 1, \qquad \frac{HG}{EG} \cdot \frac{EA}{CA} \cdot \frac{CM}{HM} = 1.$$

Since $CM = BM$ and $EA = 2FA$, division of the first equation by the second yields

$$\frac{EG}{FG} = 2\frac{BA}{CA} = 2 \cdot \frac{12}{16} = \frac{3}{2}.$$

29. (C) Instead of trying to compute $(\sqrt{3} + \sqrt{2})^6$ directly, we compute something slightly larger and easier to compute, because many terms cancel; namely we compute

$$(\sqrt{3} + \sqrt{2})^6 + (\sqrt{3} - \sqrt{2})^6.$$

When $(a + b)^{2k}$ and $(a - b)^{2k}$ are expanded by the binomial theorem, their even-powered terms are identical, and their odd-powered terms differ only in sign; so their sum is

$$2\left[a^{2k} + \binom{2k}{2}a^{2k-2}b^2 + \cdots + b^{2k}\right].$$

[†]Menelaus's theorem: If points X, Y, Z on the sides BC, CA, AB (suitably extended) of $\triangle ABC$ are collinear, then

$$\frac{BX}{CX}\frac{CY}{AY}\frac{AZ}{BZ} = 1.$$

Conversely, if this equation holds for points X, Y, Z on the three sides, then these three points are collinear.

For a proof, see e.g. H.S.M. Coxeter and S.L. Greitzer *Geometry Revisited*, vol. 19, p. 66 in this NML series.

This principle, applied with $a = \sqrt{3}$, $b = \sqrt{2}$, $k = 3$ yields

$$(\sqrt{3} + \sqrt{2})^6 + (\sqrt{3} - \sqrt{2})^6 = 2[27 + 15(18 + 12) + 8] = 970.$$

Since $0 < \sqrt{3} - \sqrt{2} < 1$, 970 is the smallest integer larger than $(\sqrt{3} + \sqrt{2})^6$.

30. (B) Let $w = \cos 36°$ and let $y = \cos 72°$. Applying the identities

$$\cos 2\theta = 2\cos^2\theta - 1 \quad \text{and} \quad \cos 2\theta = 1 - 2\sin^2\theta,$$

with $\theta = 36°$ in the first identity and $\theta = 18°$ in the second yields

$$y = 2w^2 - 1 \quad \text{and} \quad w = 1 - 2y^2.$$

Adding these last two equations yields

$$w + y = 2(w^2 - y^2) = 2(w - y)(w + y)$$

and division by $w + y$ yields $2(w - y) = 1$, so

$$x = w - y = \tfrac{1}{2}.$$

1976 Solutions

1. (B) $1 - \dfrac{1}{1-x} = \dfrac{1}{1-x}$; $1 - x - 1 = 1$; $x = -1$.

2. (B) If $x + 1 \neq 0$ then $-(x+1)^2 < 0$ and $\sqrt{-(x+1)^2}$ is not real; if $x + 1 = 0$ then $\sqrt{-(x+1)^2} = 0$. Thus $x = -1$ is the only value of x for which the given expression is real.

3. (E) The distance to each of the two closer midpoints is one; the distance to each of the other midpoints is $\sqrt{1^2 + 2^2}$.

4. (C) The sum of the terms in the new progression is
$$1 + \frac{1}{r} + \cdots + \frac{1}{r^{n-1}} = \frac{r^{n-1} + r^{n-2} + \cdots + 1}{r^{n-1}} = \frac{s}{r^{n-1}}.$$
Note: If $r = 1$, then $s = n$ and the sum of the reciprocal progression is also n. This eliminates all choices except (C).

5. (C) Let t and u be the tens' digit and units' digit, respectively, of a number which is increased by nine when its digits are reversed. Then $9 = (10u + t) - (10t + u) = 9(u - t)$ and $u = t + 1$. The eight solutions are $\{12, 23, \ldots, 89\}$.

6. (C) Let r be a solution of $x^2 - 3x + c = 0$ such that $-r$ is a solution of $x^2 + 3x - c = 0$. Then
$$r^2 - 3r + c = 0,$$
$$r^2 - 3r - c = 0.$$
which implies $2c = 0$. The solutions of $x^2 - 3x = 0$ are 0 and 3.

Note: The restriction to real c was not needed.

7. (E) The quantity $(1 - |x|)(1 + x)$ is positive if and only if either both factors are positive or both factors are negative. Both factors are positive if and only if $-1 < x < 1$, while both factors are negative if and only if $x < -1$.

8. (A) The points whose coordinates are integers with absolute value less than or equal to four form a 9×9 array, and 13 of these points are at distance less than or equal to two units from the origin. See figure.

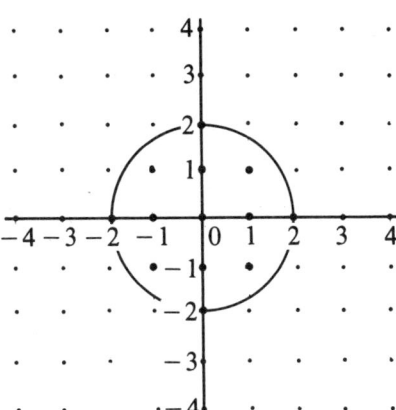

9. (D) Since F is the midpoint of BC, the altitude of $\triangle AEF$ from F to AE (extended if necessary) is one half the altitude of $\triangle ABC$ from C to AB (extended if necessary). Base AE of $\triangle AEF$ is $3/4$ of base AB of $\triangle ABC$. Therefore, the area of $\triangle AEF$ is

$$(1/2)(3/4)(96) = 36.$$

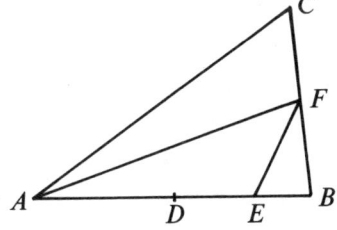

10. (D) For the given functions, $f(g(x)) = g(f(x))$ is equivalent to

$$m(px + q) + n = p(mx + n) + q,$$

which reduces to

$$mq + n = pn + q \quad \text{or} \quad n(1 - p) - q(1 - m) = 0.$$

If this last equation holds, then $f(g(x)) = g(f(x))$ is an identity, i.e. true for every value of x. If on the other hand $n(1 - p) \neq q(1 - m)$, then $f(g(x)) = g(f(x))$ has no solution in x.

Note: If the composites $f[g(x)]$ and $g[f(x)]$ of such functions f and g have the same output for some input x, then f and g commute under composition.

11. (B) The statements "P implies Q," "Not Q implies not P" and "Not P or Q" are equivalent. The given statement, statement III and statement IV are of these forms, respectively.

12. (C) There are 25 different possibilities for the number of apples a crate can contain. If there were no more than five crates containing any given number of apples, there could be at most $25(5) = 125$ crates. Since there are 128 crates, $n \geqslant 6$. We conclude that $n = 6$ by observing that it is quite possible that there are exactly six crates containing k apples in each of the cases $k = 120, 121, 122$, and exactly five crates containing k apples in each of the cases $k = 123, 124, 125, \ldots, 144$.

13. (A) If a cows give b cans in c days and A cows give B cans in C days, then we make the basic assumption

$$\frac{B}{AC} = \frac{b}{ac},$$

i.e. that the number of cans per cow-day is always the same. In the precent case $a = x$, $b = x + 1$, $c = x + 2$, and $A = x + 3$, $B = x + 5$. Thus

$$\frac{x + 5}{(x + 3)C} = \frac{x + 1}{x(x + 2)} \quad \text{and} \quad C = \frac{x(x + 2)(x + 5)}{(x + 1)(x + 3)}$$

Note: As a partial check, we verify that our answer has the appropriate units:

$$\frac{x \,\text{cows}\,(x + 2)\text{days}(x + 5)\text{cans}}{(x + 3)\text{cows}(x + 1)\text{cans}} = C \,\text{days}.$$

14. (A) Let n be the number of sides of the polygon. The sum of the interior angles of a convex polygon with n sides is $(n - 2)180°$, and the sum of n terms of an arithmetic progression is $n/2$ times the sum of the first and last terms. Therefore

$$(n - 2)180 = \frac{n}{2}(100 + 140).$$

Solving this equation for n yields $n = 6$.

15. (B) Since each of the given numbers, when divided by d, has the same remainder, d divides the differences $2312 - 1417 = 895 = 5 \cdot 179$ and $1417 - 1059 = 378 = 2 \cdot 179$; and since 179 is prime, $d = 179$. Now $1059 = 5 \cdot 179 + 164$, $r = 164$, and $d - r = 179 - 164 = 15$.

16. (E) Let G and H be the points at which the altitudes from C and F intersect sides AB and DE, respectively. Right triangles AGC and FHD are congruent, since side AG and side FH have the same length, and hypotenuse AC and hypotenuse DF have the same length. Therefore, $\angle GAC = \angle DFH$,

$$\angle ACG + \angle GAC = \angle ACG + \angle DFH = 90°,$$

so

$$\angle ACB + \angle DFE = 2\angle ACG + 2\angle DFH = 180°,$$

and

area $\triangle ABC = 2(\text{area } \triangle ACG) = 2(\text{area } \triangle DFH) = \text{area } \triangle DEF.$

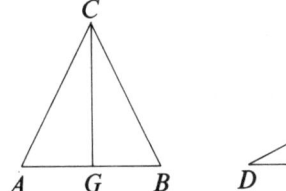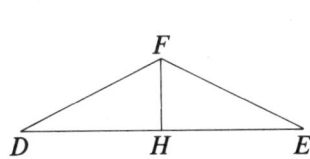

17. (A) Using trigonometric identities, we obtain

$$(\sin\theta + \cos\theta)^2 = \sin^2\theta + \cos^2\theta + 2\sin\theta\cos\theta$$
$$= 1 + \sin 2\theta$$
$$= 1 + a.$$

Since θ is acute, $\sin\theta + \cos\theta > 0$ and $\sin\theta + \cos\theta = \sqrt{1 + a}$.

18. (E) In the adjoining figure, E is the point of intersection of the circle and the extension of DB, and FG is the diameter passing through D. Let r denote the radius of the circle. Then

$$(BC)(BE) = (AB)^2,$$
$$3(DE + 6) = 36,$$
$$DE = 6.$$

Also

$$(DE)(DC) = (DF)(DG),$$
$$6 \cdot 3 = (r - 2)(r + 2),$$
$$18 = r^2 - 4,$$
$$r = \sqrt{22}.$$

19. (B) Let $r(x)$ be the desired remainder. Since its degree is less than the degree of the divisor $(x - 1)(x - 3)$, $r(x)$ is of the form $ax + b$. Thus

$$p(x) = (x - 1)(x - 3)q(x) + ax + b.$$

The given information says that $p(1) = 3$ and $p(3) = 5$. Setting $x = 1$ and then $x = 3$, we obtain

$$p(1) = a + b = 3, \qquad p(3) = 3a + b = 5,$$

so $a = 1$, $b = 2$, and $r(x) = x + 2$.

20. (E) The given equation may be written in the form

$$4(\log_a x)^2 - 8(\log_a x)(\log_b x) + 3(\log_b x)^2 = 0;$$
$$(2\log_a x - \log_b x)(2\log_a x - 3\log_b x) = 0;$$
$$\log_a x^2 = \log_b x \quad \text{or} \quad \log_a x^2 = \log_b x^3.$$

Let $r = \log_a x^2$. Then

$$a^r = x^2 \quad \text{and} \quad b^r = x, \quad \text{or} \quad a^r = x^2 \quad \text{and} \quad b^r = x^3;$$
$$a^r = b^{2r} \qquad\qquad\qquad \text{or} \quad a^{3r} = b^{2r}.$$

Since $x \neq 1$ we have $r \neq 0$; and

$$a = b^2 \quad \text{or} \quad a^3 = b^2.$$

21. (B) We recall that $1 + 3 + 5 + \cdots + 2n + 1 = (n + 1)^2$ and write the product

$$2^{1/7} \cdot 2^{3/7} \cdot \ldots \cdot 2^{(2n+1)/7} = 2^{[1+3+\cdots+2n+1]/7} = 2^{(n+1)^2/7}.$$

Since $2^{10} = 1024$, we consider values of n for which $(n + 1)^2/7$ is approximately 10:

$$2^{(7+1)^2/7} = 2^{9+(1/7)} < 2^9 \cdot 2^{1/2} = (512)(1.41\ldots)$$

$$< 1000 < 1024 = 2^{10} < 2^{(9+1)^2/7},$$

and $n = 9$.

22. (A) Let point P have coordinates (x, y) in the coordinate system in which the vertices of the equilateral triangle are $(0, 0), (s, 0)$ and $(s/2, s\sqrt{3}/2)$. Then P belongs to the locus

if and only if

$$a = x^2 + y^2 + (x - s)^2 + y^2 + (x - s/2)^2 + \left(y - s\sqrt{3}/2\right)^2,$$

or, equivalently, if and only if

$$a = \left(3x^2 - 3sx\right) + \left(3y^2 - s\sqrt{3}\,y\right) + 2s^2,$$

$$\frac{a - 2s^2}{3} = (x - s/2)^2 + \left(y - s\sqrt{3}/6\right)^2 - s^2/3,$$

$$\frac{a - s^2}{3} = (x - s/2)^2 + \left(y - s\sqrt{3}/6\right)^2.$$

Thus the locus is the empty set if $a < s^2$; the locus is a single point if $a = s^2$; and the locus is a circle if $a > s^2$.

23. (A) Since all binomial coefficients $\binom{n}{k}$ are integers, the quantity

$$\frac{n - 2k - 1}{k + 1}\binom{n}{k} = \frac{(n + 1) - 2(k + 1)}{k + 1}\binom{n}{k}$$

$$= \left(\frac{n + 1}{k + 1} - 2\right)\binom{n}{k}$$

$$= \frac{n + 1}{k + 1}\frac{n!}{k!(n - k)!} - 2\binom{n}{k}$$

$$= \frac{(n + 1)!}{(k + 1)!(n - k)!} - 2\binom{n}{k}$$

$$= \binom{n + 1}{k + 1} - 2\binom{n}{k}$$

is always an integer.

24. (C) In the adjoining figure, MF is parallel to AB and intersects KL at F. Let r, $s(= r/2)$ and t be the radii of the circles with centers K, L and M, respectively. Applying the Pythagorean theorem to $\triangle FLM$ and $\triangle FKM$ yields

$$(MF)^2 = \left(\frac{r}{2} + t\right)^2 - \left(\frac{r}{2} - t\right)^2,$$

$$(MF)^2 = (r - t)^2 - t^2.$$

Equating the right sides of these equalities yields $r/t = 4$. Therefore the desired ratio is 16.

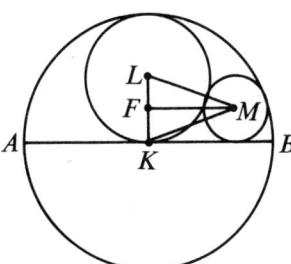

25. (D) For all positive integers n,

$$\Delta^1(u_n) = (n+1)^3 + (n+1) - n^3 - n = 3n^2 + 3n + 2$$

$$\Delta^2(u_n) = 3(n+1)^2 + 3(n+1) + 2 - 3n^2 - 3n - 2 = 6n + 6$$

$$\Delta^3(u_n) = 6$$

$$\Delta^4(u_n) = 0.$$

Note. If u_n is a polynomial in n of degree r, then $\Delta^1 u_n$ is a polynomial of degree $r-1$, $\Delta^2 u_n$ is a polynomial of degree $r-2$, etc. From this we see that $\Delta^r u_n$ is a non-zero constant sequence, and $\Delta^{r+1} u_n = 0$. In the present example $r = 3$, so $\Delta^3 u_n$ is a non-zero constant, while $\Delta^4 u_n = 0$.

Readers familiar with calculus will note an analogy with the fact that if $f(x)$ is a polynomial of degree r, then the rth derivative $D^r f(x)$ is a non-zero constant, while $D^{r+1} f(x) = 0$. The finite difference operator Δ is known as the forward difference operator and plays an important role in numerical analysis.

26. (C) In the adjoining figure, X, Y, V and W are the points of tangency of the external common tangents; and R and S are the points of tangency of the internal common tangent. From the fact that the tangents to a circle from an external point are equal, we obtain:

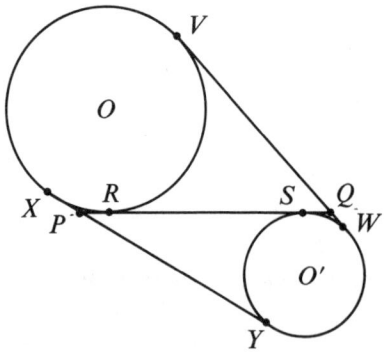

$$PR = PX, \quad PS = PY,$$
$$QS = QW, \quad QR = QV.$$

So

$$PR + PS + QS + QR = PX + PY + QW + QV,$$

and thus

$$2PQ = XY + VW.$$

Since $XY = VW$, $PQ = XY = VW.$

27. (A) A direct calculation shows that

$$\left(\frac{\sqrt{\sqrt{5} + 2} + \sqrt{\sqrt{5} - 2}}{\sqrt{\sqrt{5} + 1}} \right)^2 = 2,$$

and

$$3 - 2\sqrt{2} = 2 - 2\sqrt{2} + 1 = \left(\sqrt{2} - 1\right)^2.$$

Since a radical sign denotes the positive square root,

$$N = \sqrt{2} - \left(\sqrt{2} - 1\right) = 1.$$

28. (B) One hundred lines intersect at most at $\binom{100}{2} = \dfrac{100 \cdot 99}{2} =$ 4950 points. But the 25 lines $L_4, L_8, \ldots, L_{100}$ are parallel; hence $\binom{25}{2} = 300$ intersections are lost. Also, the 25 lines $L_1, L_5, \ldots, L_{97}$ intersect only at point A, so that $\binom{25}{2} - 1 = 299$ more intersections are lost. The maximum number of points of intersection is $4950 - 300 - 299 = 4351$.

29. (B) The table below shows the ages of Ann and Barbara at various times referred to in the problem. The first column indicates their present ages. The second column shows their ages when Barbara was half as old as Ann is now—that was $y - (x/2)$ years ago, hence Ann's age was $x - \left(y - \dfrac{x}{2}\right)$ or $\dfrac{3x}{2} - y$. The third column refers to the time when Barbara was as old as Ann had been when Barbara was half as old as Ann is—that was $y - \left(\dfrac{3x}{2} - y\right)$ or $2y - \dfrac{3x}{2}$ years ago; hence Ann's age then was $x - \left(2y - \dfrac{3x}{2}\right)$ or $\dfrac{5x}{2} - 2y$.

Ann	x	$\dfrac{3x}{2} - y$	$\dfrac{5x}{2} - 2y$
Barbara	y	$\dfrac{x}{2}$	$\dfrac{3x}{2} - y$

By the conditions stated in the problem, $x + y = 44$ and $y = \dfrac{5x}{2} - 2y$; the simultaneous solution of these yields $x = 24$.

30. (E) We observe that we can find a system of symmetric equations by the change of variables

(1) $$x = 2u, \qquad y = v, \qquad z = \frac{1}{2}w.$$

This substitution yields the transformed system

(2)
$$\begin{aligned} u + v + w &= 6, \\ uv + vw + uw &= 11, \\ uvw &= 6. \end{aligned}$$

Consider the polynomial $p(t) = (t - u)(t - v)(t - w)$, where (u, v, w) is a solution of the system (2). Then

(3) $$p(t) = t^3 - 6t^2 + 11t - 6,$$

and u, v, w are the solutions of $p(t) = 0$. Conversely, if the roots of $p(t) = 0$ are listed as a triple in any order, this triple is a solution to system (2).

It is not hard to see that $p(t) = 0$ has three distinct solutions; in fact, $p(t) = (t - 1)(t - 2)(t - 3)$. So the triple $(1, 2, 3)$ and each of its permutations satisfies the system (2). Since the change of variables (1) is one-to-one, the original system has 6 distinct solutions (x, y, z): $(2, 3, 1)$, $(2, 2, \frac{3}{2})$, $(4, 1, \frac{3}{2})$, $(4, 3, \frac{1}{2})$, $(6, 1, 1)$ or $(6, 2, \frac{1}{2})$.

Note: There are methods for determining all solutions of a system of linear equations in n unknowns, on the one hand, and of a single n-th degree polynomial in one variable, on the other. No such simple methods are generally applicable to hybrid systems of the type presented in Problem 30. The problem shows that some very special systems can be transformed into a system of n equations ($n = 3$ in the present case) involving the elementary symmetric functions of n variables, thus permitting solutions via a single polynomial equation of degree n in one variable.

1977 Solutions

1. (D) $x + y + z = x + 2x + 2y = x + 2x + 4x = 7x$.

2. (D) If three equal sides of one equilateral triangle have length s, and those of another have length t, then the triangles are congruent if and only if $s = t$.

3. (E) Let n be the number of coins the man has of each type; their total value, in cents, is

$$1 \cdot n + 5 \cdot n + 10 \cdot n + 25 \cdot n + 50 \cdot n = 91n = 273,$$

and $n = 3$; three each of five types of coins is 15 coins.

4. (C) Since the base angles of an isosceles triangle are equal, $\angle B = \angle C = 50°$,

$$\angle EDC = \angle CED = 65° \quad \text{and} \quad \angle BDF = \angle DFB = 65°.$$

It follows that

$$\angle FDE = 180° - 2(65°) = 50°.$$

Note: The measure of $\angle EDF$ is 50° even if $AB \neq AC$; see the Figure below.

$$\angle FDE = 180° - \angle EDC - \angle BDF$$
$$= 180° - \tfrac{1}{2}(180° - \angle C)$$
$$\qquad - \tfrac{1}{2}(180° - \angle B)$$
$$= \tfrac{1}{2}(\angle B + \angle C)$$
$$= \tfrac{1}{2}(180° - \angle A)$$
$$= 50°.$$

5. (A) If P is on line segment AB, then $AP + PB = AB$; otherwise $AP + PB > AB$.

6. (D) $\left(2x + \dfrac{y}{2}\right)^{-1}\left[(2x)^{-1} + \left(\dfrac{y}{2}\right)^{-1}\right] = \left(\dfrac{4x + y}{2}\right)^{-1}\left[\dfrac{1}{2x} + \dfrac{2}{y}\right]$

$$= \frac{2}{4x + y} \cdot \frac{4x + y}{2xy} = \frac{1}{xy} = (xy)^{-1}.$$

Note: The function $f(x, y) = \left(2x + \dfrac{y}{2}\right)^{-1}\left[(2x)^{-1} + \left(\dfrac{y}{2}\right)^{-1}\right]$ is *homogeneous of degree* -2, which means that $f(tx, ty) = t^{-2}f(x, y)$. Only choice D displays a function with this property.

7. (E) $\dfrac{1}{1-\sqrt[4]{2}} = \dfrac{1}{1-\sqrt[4]{2}} \cdot \dfrac{1+\sqrt[4]{2}}{1+\sqrt[4]{2}} = \dfrac{1+\sqrt[4]{2}}{1-\sqrt{2}} \cdot \dfrac{1+\sqrt{2}}{1+\sqrt{2}}$

$$= -\left(1+\sqrt[4]{2}\right)(1+\sqrt{2}).$$

8. (B) If a, b and c are all positive (negative), then 4 (respectively, -4) is formed; otherwise 0 is formed.

9. (B) Let $\overset{\frown}{AB} = x°$ and $\overset{\frown}{AD} = y°$. Then

$$3x + y = 360,$$

$$\tfrac{1}{2}(x - y) = 40.$$

Solving this pair of equations for y, we obtain $y = 30°$, and hence $\angle C = \tfrac{1}{2}y = 15°$.

10. (E) The sum of the coefficients of a polynomial $p(x)$ is equal to $p(1)$. For the given polynomial this is $(3 \cdot 1 - 1)^7 = 128$.

11. (B) If $n \leqslant x < n + 1$, then $n + 1 \leqslant x + 1 < n + 2$. Hence $[x + 1] = [x] + 1$. Choosing $x = y = 2.5$ shows that II and III are false.

12. (D) Let $a, b,$ and c denote the ages of Al, Bob and Carl, respectively. Then

$$a = 16 + (b + c) \quad \text{and} \quad a^2 = 1632 + (b + c)^2;$$

so $a^2 = 1632 + (a - 16)^2$ which yields

$$1632 - 2 \cdot 16a + 16^2 = 0 \quad \text{and} \quad a = 59.$$

Then $b + c = 59 - 16 = 43$ and $a + b + c = 102$.

OR

Since we are interested in the sum $a + b + c$, we try to write it in terms of the information supplied by the problem:

$$a + b + c = \frac{[a + b + c][a - (b + c)]}{a - (b + c)} = \frac{a^2 - (b + c)^2}{a - (b + c)}$$

$$= \frac{1632}{16} = 102.$$

13. (E) The second through the fifth terms of $\{a_n\}$ are

$$a_2, a_1a_2, a_1a_2^2, a_1^2a_2^3.$$

If these terms are in geometric progression, then the ratios of successive terms must be equal: $a_1 = a_2 = a_1a_2$. Since a_1 and a_2 are positive, it is necessary that $a_1 = a_2 = 1$. Conversely, if $a_1 = a_2 = 1$, then $\{a_n\}$ is the geometric progression $1, 1, 1, \ldots$.

14. (B) If $m + n = mn$, then

$$m + n - mn = 0, \qquad m + n(1 - m) = 0, \qquad n = \frac{m}{m - 1},$$

for $m \neq 1$. There are no solutions for which $m = 1$. The solutions $\left(m, \dfrac{m}{m - 1}\right)$ are pairs of integers only if m is 0 or 2.

OR

If $mn = m + n$, then $(m - 1)(n - 1) = 1$. Hence either $m - 1 = n - 1 = 1$ or $m - 1 = n - 1 = -1$. Thus $(m, n) = (2, 2)$ or $(0, 0)$.

15. (D) In the adjoining figure, PB and QC are radii drawn to common tangent AD of circle P and circle Q. Since $\angle PAB = \angle QDC = 30°$, we have

$$AB = CD = 3\sqrt{3}.$$

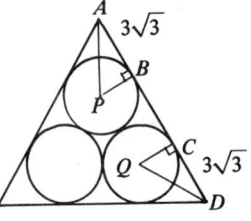

Moreover $BC = PQ = 6$, and hence $AD = 6 + 6\sqrt{3}$. Therefore the perimeter is $18 + 18\sqrt{3}$.

16. (D) Since

$$i^{n+2}\cos(45 + 90(n + 2))° = -i^n(-\cos(45 + 90n)°)$$
$$= i^n(\cos(45 + 90n)°),$$

every other term has the same value. The first is $\sqrt{2}/2$, and there are 21 terms with this value ($n = 0, 2, 4, \ldots, 40$). The second term is $i\cos 135° = -i\sqrt{2}/2$, and there are 20 terms with this value ($n = 1, 3, \ldots, 39$). Thus the sum is

$$\frac{\sqrt{2}}{2}(21 - 20i).$$

17. (B) The successful outcomes of the toss are the permutations of $(1, 2, 3)$, of $(2, 3, 4)$, of $(3, 4, 5)$ and of $(4, 5, 6)$. The probability that one of these outcomes will occur is $\dfrac{6 \cdot 4}{6^3} = \dfrac{1}{9}$.

18. (B) Let y be the desired product. By definition of logarithms, $2^{\log_2 3} = 3$. Raising both sides of this equation to the $\log_3 4$ power yields $2^{(\log_2 3)(\log_3 4)} = 4$. Continuing in this fashion, one obtains $2^y = 32$; thus $y = 5$.

OR

Express each logarithm in the problem in terms of some fixed base, say 2 (see footnote on p. 82). Then we have

$$y = \frac{\log_2 3}{\log_2 2} \cdot \frac{\log_2 4}{\log_2 3} \cdot \frac{\log_2 5}{\log_2 4} \cdots \cdot \frac{\log_2 31}{\log_2 30} \cdot \frac{\log_2 32}{\log_2 31}$$

which telescopes to

$$y = \frac{\log_2 32}{\log_2 2} = \frac{5}{1} = 5.$$

19. (A) The center of a circle circumscribing a triangle is the point of intersection of the perpendicular bisectors of the sides of the triangle. Therefore, P, Q, R and S are the intersections of the perpendicular bisectors of line segments AE, BE, CE and DE. Since line segments perpendicular to the same line are parallel, $PQRS$ is a parallelogram.

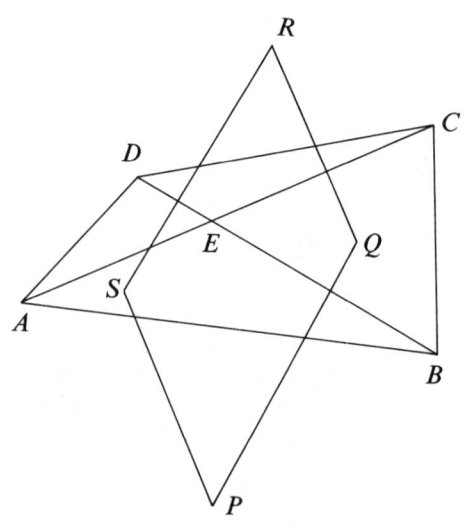

20. (E) All admissible paths end at the center "T" in the bottom row of the diagram. Our count is easier if we go from the end to the beginning of each path; that is, if we spell TSETNOC, starting at the bottom center and traversing sequences of horizontal and/or vertically upward directed segments. The count becomes easier still if we take advantage of the symmetry of the figure and distinguish those paths whose horizontal segments are directed to the left (see figure) from those whose horizontal segments are directed to the right. These two sets have the central vertical column in common and contain an equal number of paths. Starting at the bottom corner "T" in our figure, we have at each stage of the spelling the two choices of taking the next letter from above or from the left. Since there are 6 steps, this leads to 2^6 paths in this configuration. We get 2^6 paths also in the symmetric configuration. Since we have counted the central column twice, there are altogether $2 \cdot 2^6 - 1 = 127$ distinct paths.

$$C$$
$$CO$$
$$CON$$
$$CONT$$
$$CONTE$$
$$CONTES$$
$$CONTEST$$

21. (B) Subtracting the second given equation from the first yields
$$ax + x + (1 + a) = 0$$
or, equivalently,
$$(a + 1)(x + 1) = 0.$$
Hence, $a = -1$ or $x = -1$. If $a = -1$, then the given equations are identical and have (two complex but) no real solutions; $x = -1$ is a common solution to the given equations if and only if $a = 2$. Therefore, 2 is the only value of a for which the given equations have a common real solution.

22. (C) Choosing $a = b = 0$ yields
$$2f(0) = 4f(0),$$
$$f(0) = 0.$$
Choosing $a = 0$ and $b = x$ yields
$$f(x) + f(-x) = 2f(0) + 2f(x),$$
$$f(-x) = f(x).$$
Note: It can be shown that a continuous function f satisfies the given functional equation if and only if $f(x) = cx^2$, where c is some constant.

23. (B) Let a and b be the solutions of $x^2 + mx + n = 0$; then
$$-m = a + b, \qquad -p = a^3 + b^3,$$
$$n = ab, \qquad q = a^3b^3.$$
Since
$$(a + b)^3 = a^3 + 3a^2b + 3ab^2 + b^3 = a^3 + b^3 + 3ab(a + b),$$
we obtain
$$-m^3 = -p + 3n(-m) \qquad \text{or} \qquad p = m^3 - 3mn.$$

24. (D) We use the identity $\dfrac{1}{n} - \dfrac{1}{n + 2} = \dfrac{2}{n(n + 2)}$ to write the given sum in the form
$$\frac{1}{2}\left[\frac{1}{1} - \frac{1}{3} + \frac{1}{3} - \frac{1}{5} + \cdots + \frac{1}{255} - \frac{1}{257}\right],$$
which telescopes to
$$\frac{1}{2}\left[1 - \frac{1}{257}\right] = \frac{1}{2}\frac{256}{257} = \frac{128}{257},$$

OR

Use the identity
$$\frac{1}{n(n - k)} + \frac{1}{n(n + k)} = \frac{2}{(n - k)(n + k)}$$
to group successive pairs of terms in the sum. Thus letting $k = 2$, write
$$S = \frac{1}{3 \cdot 1} + \frac{1}{3 \cdot 5} + \frac{1}{7 \cdot 5} + \frac{1}{7 \cdot 9} + \cdots$$
$$= 2\left(\frac{1}{1 \cdot 5} + \frac{1}{5 \cdot 9} + \cdots\right);$$
then let $k = 4$ and write
$$S = 2 \cdot 2\left(\frac{1}{1 \cdot 9} + \frac{1}{9 \cdot 17} + \cdots\right).$$
This process will lead to $2^7 \dfrac{1}{1 \cdot 257} = \dfrac{128}{257}$.

25. (E) Let $2^k 3^l 5^m \cdots$ be the factorization of 1005! into powers of distinct primes; then n is the minimum of k and m. Now 201 of the integers between 5 and 1005 are divisible by 5; forty of these 201 integers are divisible by 5^2; eight of these forty integers are divisible by 5^3; and one of these eight integers is divisible by 5^4. Since 502 of the numbers between

2 and 1005 are even, $k > 502$; so

$$n = m = 201 + 40 + 8 + 1 = 250.$$

26. (B) If $MNPQ$ is convex, as in Figure 1, then A is the sum of areas of the triangles into which $MNPQ$ is divided by diagonal MP, so that

$$A = \tfrac{1}{2}ab \sin N + \tfrac{1}{2}cd \sin Q.$$

Similarly, dividing $MNPQ$ with diagonal NQ yields

$$A = \tfrac{1}{2}ad \sin M + \tfrac{1}{2}bc \sin P.$$

We show below that these two equations for A hold also if $MNPQ$ is not convex. Therefore, in any case,

$$A \leqslant \tfrac{1}{4}(ab + cd + ad + bc) = \frac{a+c}{2} \cdot \frac{b+d}{2}.$$

The inequality is an equality if and only if

$$\sin M = \sin N = \sin P = \sin Q = 1,$$

i.e. if and only if $MNPQ$ is a rectangle.

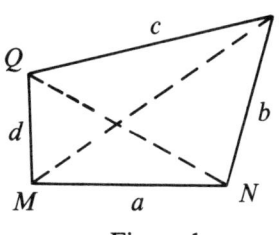

Figure 1

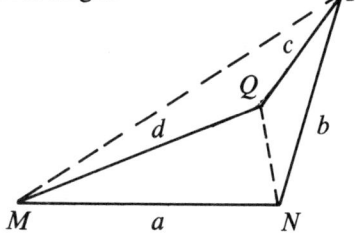

Figure 2

If $MNPQ$ is not convex, for example if interior angle[†] Q of quadrilateral $MNPQ$ is greater than $180°$ as shown in Figure 2, then A is the difference

area of $\triangle MNP$ − area of $\triangle MQP$

so that

$$A = \frac{1}{2} ab \sin N - \frac{1}{2} cd \sin \angle PQM$$

$$= \frac{1}{2} ab \sin N - \frac{1}{2} cd \sin(360° - \angle MQP)$$

$$= \frac{1}{2} ab \sin N + \frac{1}{2} cd \sin \angle MQP.$$

[†]To distinguish interior angle Q of quadrilateral $MNPQ$ from interior angle Q of $\triangle PMQ$ in Figure 2, we label the angles in the counter-clockwise direction: $\angle MQP$ is the angle through which line segment MQ must be rotated counter-clockwise about point Q to coincide with the line through P and Q. Note that

$$\angle MQP + \angle PQM = 360°.$$

27. (C) In the corner of a room two walls meet in a vertical line, and each wall meets the floor in a horizontal line. Consider these three mutually orthogonal rays as the positive axes of an x, y, z-coordinate system. A sphere with radius a, tangent to all three coordinate planes, has an equation of the form

$$(x - a)^2 + (y - a)^2 + (z - a)^2 = a^2.$$

If the point $(5, 5, 10)$ lies on such a sphere, then

$$(5 - a)^2 + (5 - a)^2 + (10 - a)^2 = a^2,$$
$$2a^2 - 40a + 150 = 0,$$
$$a^2 - 20a + 75 = 0.$$

The two solutions of this quadratic equation give the radii of spheres satisfying the given conditions. The sum of the solutions (the negative of the coefficient of a) is 20, so the sum of the diameters of the spheres is 40. (Since the last equation is equivalent to $(a - 15)(a - 5) = 0$ we see that the spheres have radii 15 and 5.)

28. (A) We shall use the identity

$$(x - 1)(x^n + x^{n-1} + \cdots + x + 1) = x^{n+1} - 1.$$

Thus, for example, $(x - 1)g(x) = x^6 - 1$. By definition of the function g, we have

$$g(x^{12}) = (x^{12})^5 + (x^{12})^4 + (x^{12})^3 + (x^{12})^2 + x^{12} + 1$$
$$= (x^6)^{10} + (x^6)^8 + (x^6)^6 + (x^6)^4 + (x^6)^2 + 1.$$

Subtracting 1 from each term on the right yields the equation

$$g(x^{12}) - 6 = \left[(x^6)^{10} - 1\right] + \left[(x^6)^8 - 1\right] + \cdots + \left[(x^6)^2 - 1\right].$$

Each expression on the right is divisible by $x^6 - 1$. We may therefore write

$$g(x^{12}) - 6 = (x^6 - 1)P(x),$$

where $P(x)$ is a polynomial in x^6. Expressing $x^6 - 1$ in terms of $g(x)$, we arrive at

$$g(x^{12}) = (x - 1)g(x)P(x) + 6.$$

When this is divided by $g(x)$, the remainder is 6.

OR

Write $g(x^{12}) = g(x)Q(x) + R(x)$, where $Q(x)$ is a polynomial and $R(x)$ is the remainder we are seeking. Since the degree of the remainder is less than that of the divisor, we know that the degree of $R(x)$ is at most 4.

Since $g(x)(x - 1) = x^6 - 1$, the five zeros of $g(x)$ are -1 and the other four (complex) sixth roots of unity; so if α is a zero of $g(x)$, then $\alpha^6 = 1$. Therefore

$$g(\alpha^{12}) = g\left[(\alpha^6)^2\right] = g(1) = 6.$$

On the other hand,

$$g(\alpha^{12}) = g(\alpha)Q(\alpha) + R(\alpha),$$
$$6 = R(\alpha),$$

and this holds for five distinct values of α. But the polynomial $R(x) - 6$ of degree less than 5 can vanish at 5 places only if $R(x) - 6 = 0$ for all x, i.e. if $R(x) = 6$ for all x.

29. (B) Let $a = x^2$, $b = y^2$ and $c = z^2$. Noting that $(a - b)^2 \geqslant 0$ implies $a^2 + b^2 \geqslant 2ab$, we see that

$$\begin{aligned}
(a + b + c)^2 &= a^2 + b^2 + c^2 + 2ab + 2ac + 2bc \\
&\leqslant a^2 + b^2 + c^2 + (a^2 + b^2) \\
&\quad + (a^2 + c^2) + (b^2 + c^2) \\
&= 3(a^2 + b^2 + c^2).
\end{aligned}$$

Therefore $n \leqslant 3$. Choosing $a = b = c > 0$ shows n is not less than three.

30. (A) In the adjoining figure, sides PQ and TS of the regular nonagon have been extended to meet at R and the circumscribed circle has been drawn. Each side of the nonagon subtends an arc of $360°/9 = 40°$; therefore

$$\angle TPQ = \angle STP$$

$$= \frac{1}{2} \cdot 3 \cdot 40° = 60°.$$

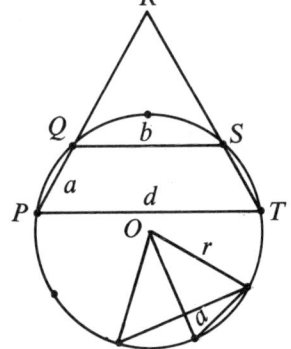

Since $QS \| PT$, it follows that both $\triangle PRT$ and $\triangle QRS$ are equilateral. Hence

$$d = PT = PR = PQ + QR = PQ + QS = a + b.$$

OR

Inscribe the nonagon in a circle of radius r. Since the chords of lengths a, b, d subtend central angles of $40°, 80°, 160°$ respectively, we have

$$a = 2r\sin 20°, \quad b = 2r\sin 40°, \quad d = 2r\sin 80°.$$

By means of the identity

$$\sin x + \sin y = 2\sin\frac{x+y}{2}\cos\frac{x-y}{2}$$

with

$$x = 40°, \quad y = 20°,$$

we obtain $\sin 40° + \sin 20° = 2\sin 30° \cos 10°$. Since $\sin 30° = \frac{1}{2}$, this yields

$$\sin 40° + \sin 20° = \cos 10° = \sin 80°,$$

and therefore $b + a = d$.

1978 Solutions

1. (B) $1 - \dfrac{4}{x} + \dfrac{4}{x^2} = \left(1 - \dfrac{2}{x}\right)^2 = 0;$ $\dfrac{2}{x} = 1.$

2. (C) Let r be the radius of the circle. Then its diameter, circumference, and area are $2r$, $2\pi r$ and πr^2, respectively. The given information reads $\dfrac{4}{2\pi r} = 2r$. Therefore $4 = 4\pi r^2$, and the area is $\pi r^2 = 1$.

3. (D) $\left(x - \dfrac{1}{x}\right)\left(y + \dfrac{1}{y}\right) = (x - y)(x + y) = x^2 - y^2.$

4. (B) $(a + b + c - d) + (a + b - c + d) + (a - b + c + d)$
$+(-a + b + c + d) = 2(a + b + c + d) = 2222.$

5. (C) Let w, x, y and z be the amounts paid by the first, second, third and fourth boy, respectively. Then since

$$w + x + y + z = 60,$$

$$w = \frac{1}{2}(x + y + z) = \frac{1}{2}(60 - w), \qquad w = 20;$$

$$x = \frac{1}{3}(w + y + z) = \frac{1}{3}(60 - x), \qquad x = 15;$$

$$y = \frac{1}{4}(w + x + z) = \frac{1}{4}(60 - y), \qquad y = 12.$$

Any of these equations now yields $z = 13$.

6. (E) If $y \neq 0$, the second equation implies $x = \frac{1}{2}$, and the first equation then implies $y = \pm \frac{1}{2}$. If $y = 0$, the first equation implies $x = 0$ or 1. Thus we have the four distinct solution pairs $(\frac{1}{2}, \frac{1}{2})$, $(\frac{1}{2}, -\frac{1}{2})$, $(0, 0)$, $(1, 0)$.

7. (E) In the accompanying figure vertices A_1 and A_3 of the hexagon lie on parallel sides twelve inches apart, and M is the midpoint of A_1A_3. Thus $\triangle A_1A_2M$ is a 30°–60°–90° triangle, and

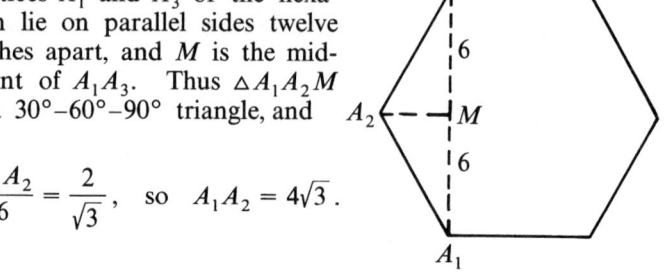

$$\frac{A_1A_2}{6} = \frac{2}{\sqrt{3}}, \quad \text{so} \quad A_1A_2 = 4\sqrt{3}.$$

8. (D) In the first sequence, it takes three equal steps to get from x to y; in the second, it takes 4. Hence the ratio of these step sizes is

$$\frac{a_2 - a_1}{b_2 - b_1} = \frac{(y - x)/3}{(y - x)/4} = \frac{4}{3}.$$

9. (B) $\left| x - \sqrt{(x-1)^2} \right| = |x - |x - 1|\,| = |x - (1 - x)|$
$= |-1 + 2x| = 1 - 2x.$

10. (B) For each point A other than P, the point of intersection of circle C with the ray beginning at P and passing through A is the point on circle C closest to A. Therefore the ray beginning at P and passing through B is the set of all points A such that B is the point on circle C which is closest to point A.

11. (C) If the line with equation $x + y = r$ is tangent to the circle with equation $x^2 + y^2 = r$, then the distance between the point of tangency $\left(\frac{r}{2}, \frac{r}{2}\right)$ and the origin is $\sqrt{r}$. Therefore, $\left(\frac{r}{2}\right)^2 + \left(\frac{r}{2}\right)^2 = r$ and $r = 2$.

12. (E) Let $x = \angle BAC = \angle BCA$; $y = \angle CBD = \angle CDB$ and $z = \angle DCE = \angle DEC$. Applying the theorem on exterior angles to $\triangle ABC$ and $\triangle ACD$ and the theorem on the sum of the interior angles of a triangle to $\triangle ADE$ yields

$$y = 2x,$$

$$z = x + y = 3x,$$

$$x + \angle ADE + z = 180°,$$

$$140° + 4x = 180°,$$

$$x = 10°.$$

13. (B) Since the constant term of a monic[†] quadratic equation is the product of its roots,

$$b = cd, \qquad d = ab.$$

Since the coefficient of x in a monic quadratic equation is the negative of the sum of its roots,

$$-a = c + d, \qquad -c = a + b;$$

thus $a + c + d = 0 = a + b + c$, and $b = d$. But the equations $b = cd$ and $d = ab$ imply, since $b = d \neq 0$, that $1 = a = c$. Therefore, $b = d = -2$, and

$$a + b + c + d = -2.$$

14. (C) Since $n^2 - an + b = 0$, and $a = (18)_n = 1 \cdot n + 8$, we have $b = an - n^2 = (n + 8)n - n^2 = 8n = (80)_n$.

OR

The sum of the roots of $x^2 - ax + b = 0$ is $a = (18)_n$. Since one root is $n = (10)_n$, the other is $(18)_n - (10)_n = (8)_n = 8$. Hence the product b of the roots is $(8)_n \cdot (10)_n = (80)_n$.

[†]A monic polynomial is one in which the coefficient of the highest power of the variable is 1.

15. (A) If $\sin x + \cos x = \dfrac{1}{5}$, then $\cos x = \dfrac{1}{5} - \sin x$ and

$$\cos^2 x = 1 - \sin^2 x = \left(\frac{1}{5} - \sin x\right)^2;$$

so

$$25 \sin^2 x - 5 \sin x - 12 = 0.$$

The solutions of $25s^2 - 5s - 12 = 0$ are $s = 4/5$ and $s = -3/5$. Since $0 \leqslant x < \pi$, $\sin x \geqslant 0$, so $\sin x = 4/5$ and $\cos x = (1/5) - \sin x = -3/5$. Hence $\tan x = -4/3$.

16. (E) Label the people $A_1, A_2, \ldots, A_N$ in such a way that A_1 and A_2 are a pair that did not shake hands with each other. Possibly every other pair of people shook hands, so that only A_1 and A_2 did not shake with everyone else. Therefore, at most $N - 2$ people shook hands with everyone else.

17. (D) Applying the given relation with $x = 3/y$ yields

$$f\left(\frac{9 + y^2}{y^2}\right)^{2\sqrt{3/y}} = \left[f\left(\left(\frac{3}{y}\right)^2 + 1\right)^{\sqrt{3/y}}\right]^2 = k^2.$$

18. (C) We seek the smallest integer n such that $\sqrt{n} - \sqrt{n-1} < \dfrac{1}{100}$. Since, for positive a, $a < c$ if and only if $\dfrac{1}{a} > \dfrac{1}{c}$, we must find the smallest integer such that

$$\frac{1}{\sqrt{n} - \sqrt{n-1}} = \sqrt{n} + \sqrt{n-1} > 100.$$

Since $\sqrt{2500} + \sqrt{2499} < 100$ and $\sqrt{2501} + \sqrt{2500} > 100$, the least such integer is 2501.

19. (C) Since $50p + 50(3p) = 1$, $p = .005$. Since there are seven perfect squares not exceeding 50 and three greater than 50 but not exceeding 100, the probability of choosing a perfect square is $7p + 3(3p) = .08$.

20. (A) Observe that the given relations imply

$$\frac{a + b - c}{c} + 2 = \frac{a - b + c}{b} + 2 = \frac{-a + b + c}{a} + 2,$$

that is,

$$\frac{a + b + c}{c} = \frac{a + b + c}{b} = \frac{a + b + c}{a}.$$

These equalities are satisfied if $a + b + c = 0$, in which case $x = -1$. If $a + b + c \neq 0$, then dividing each member of the second set of equalities by $a + b + c$ and taking the reciprocals of the results yields $a = b = c$. In that case $x = 8$.

21. (A) Since $\log_b a = \dfrac{1}{\log_a b}$ (see footnote on p. 82), we see that

$$\frac{1}{\log_3 x} + \frac{1}{\log_4 x} + \frac{1}{\log_5 x} = \log_x 3 + \log_x 4 + \log_x 5$$

$$= \log_x 60 = \frac{1}{\log_{60} x}.$$

22. (D) Since each pair of statements on the card is contradictory, at most one of them is true. The assumption that none of the statements is true implies the fourth statement is true. Hence, there must be exactly one true statement on the card. In fact, it is easy to verify that the third statement is true.

23. (C) Let FG be an altitude of $\triangle AFB$, and let x denote the length of AG. From the adjoining figure it may be seen that

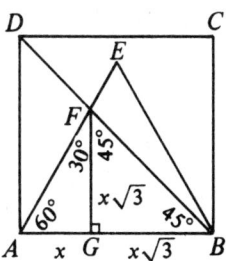

$$\sqrt{1 + \sqrt{3}} = AB = x(1 + \sqrt{3}),$$

$$1 + \sqrt{3} = x^2(1 + \sqrt{3})^2,$$

$$1 = x^2(1 + \sqrt{3}).$$

The area of $\triangle ABF$ is

$$\tfrac{1}{2}(AB)(FG) = \tfrac{1}{2}x^2(1 + \sqrt{3})\sqrt{3} = \tfrac{1}{2}\sqrt{3}.$$

24. (A) Let $a = x(y - z)$ and observe that the identity

$$x(y - z) + y(z - x) + z(x - y) = 0$$

implies

$$a + ar + ar^2 = 0;$$
$$1 + r + r^2 = 0.$$

25. (D) The boundary of the set of points satisfying the conditions is shown in the adjoining figure.

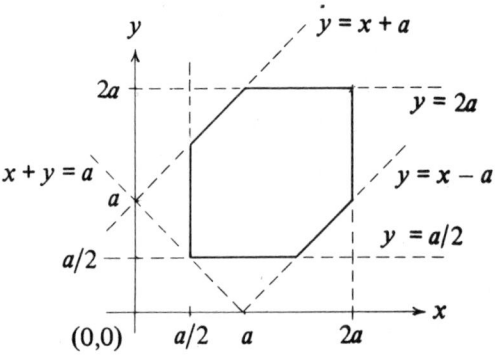

26. (B) In the first figure, K is any circle tangent to AB and passing through C. Let T be the point of tangency and NT the diameter through point T. Let CH be the altitude of $\triangle ABC$ from C. Then $NT \geqslant CT \geqslant CH$ with $NT = CH$ if and only if $N = C$, and $T = H$. There *is* such a circle with diameter CH, so it is the (unique) circle P of the problem, shown in the second figure. By the converse of the Pythagorean theorem, $\angle RCQ = 90°$; thus QR is also a diameter of P. Since $\triangle CBH \sim \triangle ABC$, $\dfrac{CH}{6} = \dfrac{8}{10}$, so $QR = CH = 4.8$.

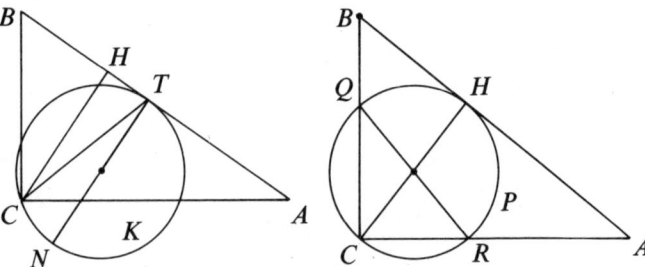

Figure 1   Figure 2

27. (C) A positive integer has a remainder of 1 when divided by any of the integers from 2 through 11 if and only if the integer is of the form $mt + 1$, where t is a nonnegative integer and $m = 2^3 \cdot 3^2 \cdot 5 \cdot 7 \cdot 11 = 27{,}720$ is the least common multiple of $2, 3, \ldots, 11$. Therefore, consecutive integers with the desired property differ by 27,720.

28. (E) Triangle $A_2 A_3 A_4$ has vertex angles $60°, 30°, 90°$, respectively. Since $\angle A_1 A_2 A_3 = 60°$, and $A_2 A_4$ and $A_2 A_5$ have the same length, $\triangle A_2 A_4 A_5$ is equilateral. Therefore, $\triangle A_3 A_4 A_5$ has vertex angles $30°, 30°, 120°$, respectively. Then $\triangle A_4 A_5 A_6$ has vertex angles $30°, 60°, 90°$, respectively. Finally, since $\angle A_4 A_5 A_6 = 60°$

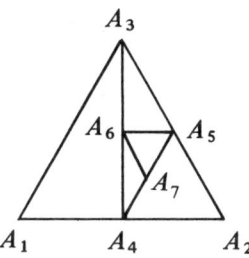

and $A_5 A_6$ and $A_5 A_7$ have the same length, $\triangle A_5 A_6 A_7$ is again equilateral. The next cycle yields four triangles, each similar to the corresponding triangle in the previous cycle. Therefore $\triangle A_n A_{n+1} A_{n+2} \sim \triangle A_{n+4} A_{n+5} A_{n+6}$ with A_n and A_{n+4} as corresponding vertices. Thus

$$\angle A_{44} A_{45} A_{43} = \angle A_4 A_5 A_3 = 120°.$$

29. (D) Since the length of base AA' of $\triangle AA'B$ is the same as the length of base AD of $\triangle ABD$, and the corresponding altitude of $\triangle AA'B'$ has twice the length of the corresponding altitude of $\triangle ABD$,

$$\text{area} \triangle AA'B' = 2\,\text{area} \triangle ADB,$$

see figure on next page. (Alternatively, we could let θ be the measure of $\angle DAB$ and observe

$$\text{area} \triangle AA'B' = \frac{1}{2}(AD)(2AB)\sin(180° - \theta)$$

$$= 2\left[\frac{1}{2}(AD)(AB)\sin\theta\right] = 2\,\text{area} \triangle ABD.)$$

Similarly

$$\text{area} \triangle BB'C' = 2\,\text{area} \triangle BAC,$$
$$\text{area} \triangle CC'D' = 2\,\text{area} \triangle CBD,$$
$$\text{area} \triangle DD'A' = 2\,\text{area} \triangle DCA.$$

Therefore

area $A'B'C'D'$

$= (\text{area } \triangle AA'B' + \text{area } \triangle BB'C')$

$\quad + (\text{area } \triangle CC'D' + \text{area } \triangle DD'A')$

$\quad + \text{area } ABCD$

$= 2(\text{area } \triangle ABD + \text{area } \triangle BAC)$

$\quad + 2(\text{area } \triangle CBD + \text{area } \triangle DCA)$

$\quad + \text{area } ABCD$

$= 5 \text{ area } ABCD = 50.$

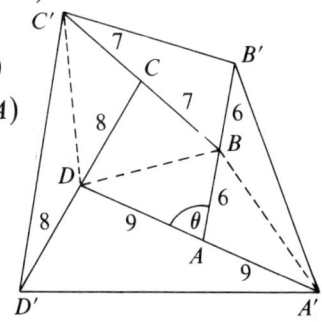

30. (E) Let $7m, 5m$ be the total number of matches won by women and men, respectively. Now there are

$$\frac{n(n-1)}{2} \text{ matches between women, hence won by women.}$$

There are

$$\frac{2n(2n-1)}{2} = n(2n-1) \text{ matches between men, hence}$$

won by men. Finally, there are

$$2n \cdot n = 2n^2 \text{ mixed matches, of which}$$

$$k = 7m - \frac{n(n-1)}{2} \text{ are won by women, and}$$

$$2n^2 - k = 5m - n(2n-1) \text{ by men.}$$

We note for later use that $k \leqslant 2n^2$. Then

$$\frac{7m}{5m} = \frac{\frac{1}{2}n(n-1) + k}{n(2n-1) + 2n^2 - k} = \frac{1}{2}\frac{n(n-1) + 2k}{4n^2 - n - k},$$

$$5n(n-1) + 10k = 14(4n^2 - n - k),$$

$$51n^2 - 9n - 24k = 3(17n^2 - 3n - 8k) = 0,$$

$$17n^2 - 3n = 8k,$$

$$\frac{17n^2 - 3n}{8} = k.$$

Since $k \leqslant 2n^2$, it follows that $17n^2 - 3n \leqslant 16n^2$, so $n(n - 3) \leqslant 0$, $n \leqslant 3$. Since k is an integer, $n \neq 1$ or 2; hence $n = 3$.

OR

The total number of games played was

$$\binom{3n}{2} = \frac{3n(3n - 1)}{2}.$$

Suppose that $7m$ games were won by women, and $5m$ games were won by men. Then

$$7m + 5m = 12m = \frac{3n(3n - 1)}{2} \quad \text{and} \quad 3n(3n - 1) = 24m.$$

Since m is an integer, $3n(3n - 1)$ is a multiple of 24. But this is not true for $n = 2, 4, 6, 7$. (It is true for $n = 3$.)

1979 Solutions

1. (D) Rectangle $DEFG$ has area 18, one quarter of the area of rectangle $ABCD$.

2. (D) $\dfrac{1}{x} - \dfrac{1}{y} = \dfrac{y - x}{xy} = -\dfrac{x - y}{xy} = -\dfrac{xy}{xy} = -1.$

3. (C) Since $\angle DAE = 90° + 60° = 150°$ and $DA = AB = AE,$ $\angle AED = \left(\dfrac{180 - 150}{2}\right)° = 15°.$

4. (E) $x[x\{x(2 - x) - 4\} + 10] + 1$
   $= x[x\{2x - x^2 - 4\} + 10] + 1 = x[2x^2 - x^3 - 4x + 10] + 1$
   $= -x^4 + 2x^3 - 4x^2 + 10x + 1.$

5. (D) The units and hundreds digits of the desired number must clearly be equal. The largest such even three digit number is 898. The sum of its digits is 25.

6. (A) Add and subtract directly or note that

   $$\frac{3}{2} + \frac{5}{4} + \cdots + \frac{65}{64} - 7$$

   $$= \left(1 + \frac{1}{2}\right) + \left(1 + \frac{1}{4}\right) + \cdots + \left(1 + \frac{1}{64}\right) - 7$$

   $$= \frac{1}{2} + \frac{1}{4} + \cdots + \frac{1}{64} - 1 = -\frac{1}{64}.$$

7. (E) Let $x = n^2, n \geqslant 0$. Then

   $$(n + 1)^2 = n^2 + 2n + 1 = x + 2\sqrt{x} + 1.$$

8. (C) The area of the smallest region bounded by $y = |x|$ and $x^2 + y^2 = 4$ is shown in the adjoining figure. Its area is

   $$\frac{1}{4}(\pi 2^2) = \pi.$$

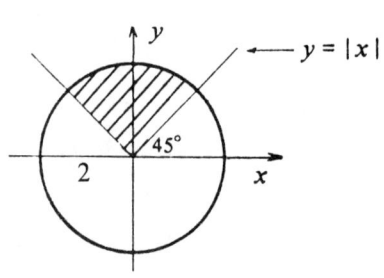

9. (E) $\sqrt[3]{4}\sqrt[4]{8} = (2^2)^{1/3}(2^3)^{1/4} = 2^{2/3}2^{3/4} = 2^{17/12} = 2^{1+(5/12)}$
$= 2 \cdot 2^{5/12} = 2\sqrt[12]{32}$.

10. (D) Let C be the center of the hexagon; then the area of $Q_1Q_2Q_3Q_4$ is the sum of the areas of the three equilateral triangles $\triangle Q_1Q_2C$, $\triangle Q_2Q_3C$, $\triangle Q_3Q_4C$, each of whose sides have length 2. The area of an equilateral triangle of side s is $\frac{1}{4}s^2\sqrt{3}$. Therefore,

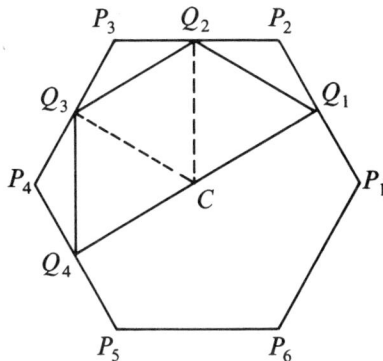

$$\text{area } Q_1Q_2Q_3Q_4 = 3\frac{2^2\sqrt{3}}{4} = 3\sqrt{3}.$$

11. (B) Summing the arithmetic progressions yields

$$\frac{115}{116} = \frac{1 + 3 + \cdots + (2n - 1)}{2 + 4 + \cdots + 2n}$$

$$= \frac{n^2}{n(n+1)} = \frac{n}{n+1};$$

hence $n = 115$.

12. (B) Draw line segment BO, and let x and y denote the measures of $\angle EOD$ and $\angle BAO$, respectively. Observe that $AB = OD = OE = OB$, and apply the theorem on exterior angles of triangles to $\triangle ABO$ and $\triangle AEO$ to obtain

$$\angle EBO = \angle BEO = 2y$$

and

$$x = 3y.$$

Thus

$$45° = 3y,$$
$$15° = y.$$

OR

Since the measure of an angle formed by two secants is half the difference of the intercepted arcs,

$$y = \frac{1}{2}(x - y), \quad y = \frac{x}{3} = \frac{45°}{3} = 15°.$$

13. (A) Consider two cases:

Case 1: If $x \geqslant 0$ then $y - x < \sqrt{x^2}$ if and only if $y - x < x$, or equivalently, $y < 2x$.

Case 2: If $x < 0$ then $y - x < \sqrt{x^2}$ if and only if $y - x < -x$, or equivalently, $y < 0$.

The pairs (x, y) satisfying the conditions described in either Case 1 or Case 2 are exactly the pairs (x, y) such that $y < 0$ or $y < 2x$ (see adjoining diagram for geometric interpretation).

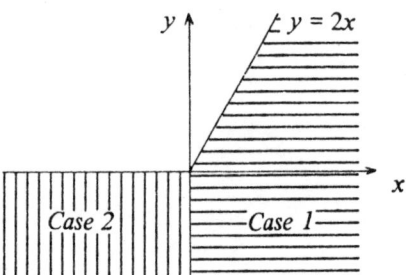

Note 1: Since $\sqrt{x^2} = |x|$, the given inequality can be written

$$y < x + |x| = \begin{cases} 2x & \text{for } x \geqslant 0 \\ 0 & \text{for } x < 0 \end{cases}$$

and the graph of $x + |x|$ is the boundary of the shaded region in the figure.

Note 2: The given inequality holds if $x = 1$, $y = 0$, and also if $x = y = -1$. This eliminates all choices except (A).

14. (C) Let a_n denote the nth number in the sequence; then

$$a_n = \frac{a_1 a_2 \cdots a_n}{a_1 a_2 \cdots a_{n-1}} = \frac{n^2}{(n-1)^2}$$

for $n \geqslant 2$. Thus, the first five numbers in the sequence are $1, 4, \frac{9}{4}, \frac{16}{9}, \frac{25}{16}$, and the desired sum is $\frac{9}{4} + \frac{25}{16} = \frac{61}{16}$.

15. (E) If each jar contains a total of x liters of solution, then one jar contains $\dfrac{px}{p+1}$ liters of alcohol and $\dfrac{x}{p+1}$ liters of water, and the other jar contains $\dfrac{qx}{q+1}$ liters of alcohol and $\dfrac{x}{q+1}$ liters of water. The ratio of the volume of alcohol to the volume of water in the mixture is then

$$\frac{\left(\dfrac{p}{p+1} + \dfrac{q}{q+1}\right)x}{\left(\dfrac{1}{p+1} + \dfrac{1}{q+1}\right)x} = \frac{p(q+1) + q(p+1)}{(q+1) + (p+1)}$$

$$= \frac{p+q+2pq}{p+q+2}.$$

16. (E) Since $A_1, A_2, A_1 + A_2$ are in arithmetic progression,
$$A_2 - A_1 = (A_1 + A_2) - A_2; \qquad 2A_1 = A_2.$$
If r is the radius of the smaller circle, then
$$9\pi = A_1 + A_2 = 3A_1 = 3\pi r^2; \qquad r = \sqrt{3}.$$

17. (C) Since the length of any side of a triangle is less than the sum of the lengths of the other sides,

$x < y - x + z - y = z - x$, which implies $x < \dfrac{z}{2}$;

$y - x < x + z - y$, which implies $y < x + \dfrac{z}{2}$;

$z - y < x + y - x$, which implies $y > \dfrac{z}{2}$.

Therefore, only statements I and II are true.

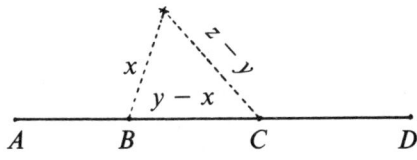

18. (C) Since $\log_b a = \dfrac{1}{\log_a b}$ (see footnote on p. 82),

$$\log_5 10 = \frac{1}{\log_{10} 5} = \frac{1}{\log_{10} 10 - \log_{10} 2} \approx \frac{1}{.699} \approx \frac{10}{7}.$$

(The value of $\log_{10} 3$ was not used.)

19. (A) Since $256^{32} = (2^8)^{32} = 2^{256}$, the given equation is equivalent to

$$\left(\frac{x}{2}\right)^{256} = 1.$$

Among the 256 256th roots of 1, only 1 and -1 are real. Thus $x = 2$ or $x = -2$ and $2^2 + (-2)^2 = 8$.

20. (C) Let $x = \text{Arctan } a$ and $y = \text{Arctan } b$:

$$(a + 1)(b + 1) = 2,$$

$$(\tan x + 1)(\tan y + 1) = 2,$$

$$\tan x + \tan y = 1 - \tan x \tan y,$$

$$\frac{\tan x + \tan y}{1 - \tan x \tan y} = 1.$$

The left hand side is $\tan(x + y)$, so

$$\tan(x + y) = 1, \quad x + y = \frac{\pi}{4} = \text{Arctan } a + \text{Arctan } b.$$

OR

Substituting $a = \frac{1}{2}$ in the equation $(a + 1)(b + 1) = 2$ and solving for b, we obtain $b = \frac{1}{3}$. Then

$$\tan(x + y) = \frac{\tan x + \tan y}{1 - \tan x \tan y} = \frac{a + b}{1 - ab} = \frac{\frac{1}{2} + \frac{1}{3}}{1 - \frac{1}{6}} = 1,$$

so $x + y = \dfrac{\pi}{4}$.

Note: $x + y$ cannot be $(\pi/4) + n\pi$ for a non-zero integer n because $0 < x, y < \pi/2$.

21. (B) In the adjoining figure, O is the center of the circle and x, y are the lengths of the legs of the triangle. So

$$h = (y - r) + (x - r)$$

$$= x + y - 2r,$$

and

$$x + y = h + 2r.$$

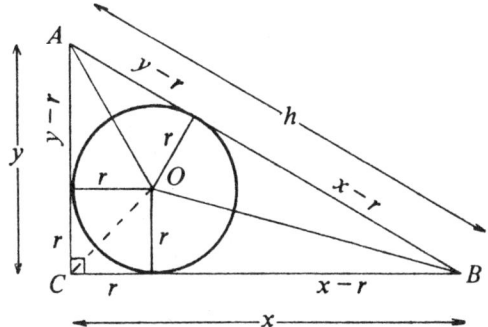

The area of $\triangle ABC$ is the sum of the areas of triangles AOB, BOC and AOC, whose altitudes have length r. Thus

$$\text{area } \triangle ABC = \frac{1}{2}(xr + yr + hr) = \frac{r}{2}(x + y + h)$$

$$= \frac{r}{2}(h + 2r + h) = r(h + r).$$

Thus the desired ratio is $\dfrac{\pi r^2}{hr + r^2} = \dfrac{\pi r}{h + r}.$

Note: Alternatively, the area of $\triangle ABC$ is

$$\frac{1}{2}xy = \frac{1}{2}\frac{(x + y)^2 - (x^2 + y^2)}{2}$$

$$= \frac{1}{4}\left[(h + 2r)^2 - h^2\right] = hr + r^2.$$

22. (A) The left member of the given equation can be factored into $m(m + 1)(m + 5)$ and rewritten in the form

$$m(m + 1)(m + 2 + 3) = m(m + 1)(m + 2) + 3m(m + 1).$$

For all integers m the first term is the product of three consecutive integers, hence divisible by 3, and the second term is obviously divisible by 3. So for all integers m, the left side is divisible by 3.

   The right side, $3[9n^3 + 3n^2 + 3n] + 1$ has remainder 1 when divided by 3. Therefore there are no integer solutions of the given equation.

23. (C) Let M and N be the midpoints of AB and CD, respectively. We claim that M and N are the unique choices for P and Q which minimize the distance PQ. To show this we consider the set S of all points equidistant from A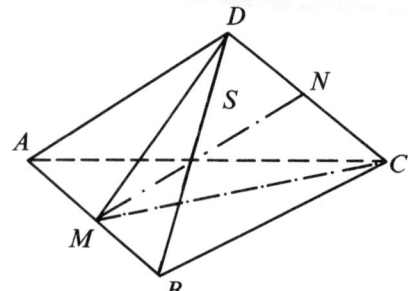

and B. S is the plane perpendicular to AB through M. Since C and D are equidistant from A and B, they lie in S, and so does the line through C and D. Now M is the foot of the perpendicular to AB from any point Q on CD. Therefore, if P is any point on AB,

$$MQ < PQ \quad \text{unless} \quad P = M.$$

Similarly, the plane through N perpendicular to CD contains AB. In particular, $MN \perp CD$; thus

$$MN < MQ \quad \text{unless} \quad Q = N.$$

By transitivity, $MN < PQ$ unless $P = M$ and $Q = N$. This proves the claim.

To compute the length of MN, we note that MN is the altitude of isosceles $\triangle DMC$ with sides of lengths $\sqrt{3}/2$, $\sqrt{3}/2$, 1. The Pythagorean theorem now yields

$$MN = \sqrt{(MC)^2 - (NC)^2} = \sqrt{\frac{3}{4} - \frac{1}{4}}$$

$$= \frac{\sqrt{2}}{2} = \text{minimal distance } PQ.$$

24. (E) Let E be the intersection of lines AB and CD, and let β and θ be the measures of $\angle EBC$ and $\angle ECB$, respectively; see figure. Since

$$\cos \beta = -\cos B = \sin C = \sin \theta,$$

$\beta + \theta = 90°$, so $\angle BEC$ is a right angle, and

$$BE = BC \sin \theta = 3; \quad CE = BC \sin \beta = 4.$$

Therefore, $AE = 7$, $DE = 24$ and AD, which is the hypotenuse of right triangle ADE, is $\sqrt{7^2 + 24^2} = 25$.

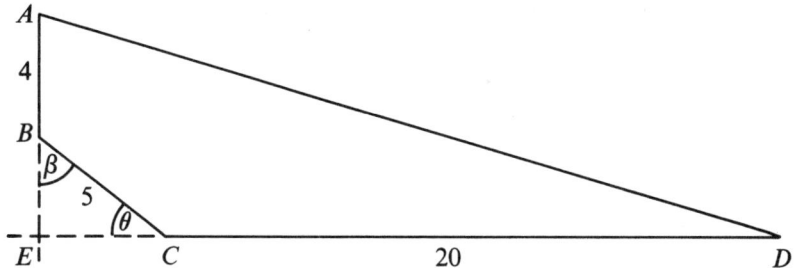

25. (B) Let $a = -\frac{1}{2}$. Applying the remainder theorem yields $r_1 = a^8$, and solving the equality $x^8 = (x - a)q_1(x) + r_1$ for $q_1(x)$ yields

$$q_1(x) = \frac{x^8 - a^8}{x - a} = x^7 + ax^6 + \cdots + a^7$$

[or, by factoring a difference of squares three times,

$$q_1(x) = (x^4 + a^4)(x^2 + a^2)(x + a)].$$

Applying the remainder theorem to determine the remainder when $q_1(x)$ is divided by $x - a$ yields

$$r_2 = q_1(a) = 8a^7 = -\frac{1}{16}.$$

Note: For solvers familiar with calculus, there is another way to find $q_1(a)$. We differentiate the identity $x^8 = (x - a)q_1(x) + r_1$ with respect to x, obtaining

$$8x^7 = q_1(x) + (x - a)q_1'(x).$$

Setting $x = a$ yields $8a^7 = q_1(a)$.

26. (B) Substitute $x = 1$ into the functional equation and solve for the first term on the right side to obtain

$$f(y + 1) = f(y) + y + 2.$$

Since $f(1) = 1$, one sees by successively substituting $y = 2, 3, 4, \ldots$ that $f(y) > 0$ for every positive integer. Therefore, for y a positive integer, $f(y + 1) > y + 2 > y + 1$, and $f(n) = n$ has no solutions for integers $n > 1$. Solving the above equation for $f(y)$ yields

$$f(y) = f(y + 1) - (y + 2).$$

Successively substituting $y = 0, -1, -2, \ldots$ into this equation yields $f(0) = -1$, $f(-1) = -2$, $f(-2) = -2$, $f(-3) = -1$, $f(-4) = 1$. Now $f(-4) > 0$ and, for $y < -4$, $-(y + 2) > 0$. Thus, for $y < -4$, $f(y) > 0$. Therefore,

$f(n) \neq n$ for $n < -4$; and the solutions $n = 1, -2$ are the only ones.

OR

We write the functional equation in the form

$$f(x + y) - f(y) = f(x) + xy + 1.$$

Setting $x = 1$ and using the given value $f(1) = 1$, we find

$$f(y + 1) - f(y) = y + 2.$$

We now set $y = 0, 1, 2, \ldots n$ successively, then $y = 0, -1, -2, \ldots - m$ successively, and obtain

$f(1) - f(0) = 2$ (so $f(0) = -1$)	$f(0) - f(-1) = 1$
$f(2) - f(1) = 3$	$f(-1) - f(-2) = 0$
$\cdots\cdots\cdots\cdots$	$\cdots\cdots\cdots\cdots\cdots$
$f(n) - f(n - 1) = n + 1$	$f(-(m - 1)) - f(-m) = -(m - 2).$

Adding the set of equations in the left column and cancelling like terms with opposite signs yields

$$f(n) - f(0) = 2 + 3 + \cdots + n + 1 = -1 + \sum_{i=1}^{n+1} i.$$

Recalling that the sum of the first k positive integers is $\frac{1}{2}k(k + 1)$ and using the value $f(0) = -1$, we obtain

$$f(n) + 1 = \frac{(n + 1)(n + 2)}{2} - 1,$$

(1) $$f(n) = \tfrac{1}{2}(n^2 + 3n - 2)$$

for each non-negative integer n.

The same procedure applied to the column of equations on the right above shows that equation (1) is valid also for negative integers; thus (1) holds for all integers.

The equation $f(n) = n$ is equivalent to

$$n^2 + 3n - 2 = 2n,$$
$$n^2 + n - 2 = (n + 2)(n - 1) = 0,$$
$$n = 1 \quad \text{or} \quad n = -2;$$

so there is only one integer solution other than $n = 1$.

Note 1: The computations involving the equations in the right column above can be avoided by setting $y = -x$ in the original equation to obtain

$$f(x) + f(-x) = f(0) + x^2 - 1,$$

or since $f(0) = -1$,

(2) $$f(x) = -f(-x) + x^2 - 2$$

for all x. Let x be negative and substitute (1) in (2) to get

$$f(x) = -\tfrac{1}{2}\big[(-x)^2 + 3(-x) - 2\big] + x^2 - 2 = \tfrac{1}{2}(x^2 + 3x - 2).$$

Thus negative integers as well as positive satisfy (1). In fact, it is easy to check that the function

$$f(x) = \frac{x^2 + 3x - 2}{2}$$

satisfies the given functional equation for all *real* x and y.

Note 2: The technique used in the second solution is frequently applied to solve many commonly occurring functional equations involving the expression $\Delta f(x) = f(x+1) - f(x)$; $\Delta f(x)$ is called the "first difference" of $f(x)$ and behaves, in many ways, like the first derivative $f'(x)$ of $f(x)$. For example

$$f'(x) = x + 2 \quad \text{implies} \quad f(x) = \frac{x^2}{2} + 2x + c;$$

$$\Delta f(x) = x + 2 \quad \text{implies} \quad f(x) = \frac{x(x-1)}{2} + 2x + c,$$

where c is a constant. The study of functions by means of $\Delta f(x)$ is called the theory of finite differences.[†]

27. (E) These six statements are equivalent for integers b, c with absolute value at most 5:

(1) the equation $x^2 + bx + c = 0$ has positive roots;
(2) the equation $x^2 + bx + c = 0$ has real roots, the smaller of which is positive;
(3) $\sqrt{b^2 - 4c}$ is real and $-b - \sqrt{b^2 - 4c} > 0$;
(4) $0 \leqslant b^2 - 4c < b^2$ and $b < 0$;
(5) $0 < c \leqslant \dfrac{b^2}{4}$ and $b < 0$;
(6) $b = -2$ and $c = 1$; or $b = -3$ and $c = 1$ or 2; or $b = -4$ and $c = 1, 2, 3$ or 4; or $b = -5$ and $c = 1, 2, 3, 4$ or 5.

The roots corresponding to the pairs (b, c) described in (6) will be distinct unless $b^2 = 4c$. Thus, deleting (b, c) = (-2, 1) and (-4, 4) from the list in (6) yields the ten pairs resulting in distinct positive roots.

The desired probability is then $1 - \dfrac{10}{11^2} = \dfrac{111}{121}$.

[†]Some references to this theory or to the theory of more general functional equations are: *Finite Differences* by S. Goldberg; "Functional Equations in Secondary Mathematics", by S. B. Maurer, an article in *The Mathematics Teacher*, April 1974.

28. (D) Denote by A' the analogous intersection point of the circles
with centers at B and C, so that, by symmetry, $A'B'C'$ and
ABC are both equilateral triangles. Again, by symmetry,
$\triangle ABC$ and $\triangle A'B'C'$ have a common centroid; call it K. Let
M be the midpoint of the line segment BC. From the trian-
gle $A'BC$ we see that the length $A'M = \sqrt{r^2 - 1}$. Since corresponding lengths in similar triangles are proportional,

$$\frac{B'C'}{BC} = \frac{A'K}{AK}.$$

Since equilateral $\triangle ABC$ has sides of length 2, we find that
$B'C' = 2\dfrac{A'K}{AK}$ and also that altitude AM has length $\sqrt{3}$.
Consequently

$$AK = \frac{2}{3}AM = \frac{2}{3}\sqrt{3}, \qquad MK = \frac{1}{3}AM = \frac{1}{3}\sqrt{3};$$

and

$$A'K = A'M + MK = \sqrt{r^2 - 1} + (\sqrt{3}/3).$$

Thus

$$B'C' = 2\frac{\sqrt{r^2 - 1} + (\sqrt{3}/3)}{2(\sqrt{3}/3)} = \sqrt{3(r^2 - 1)} + 1.$$

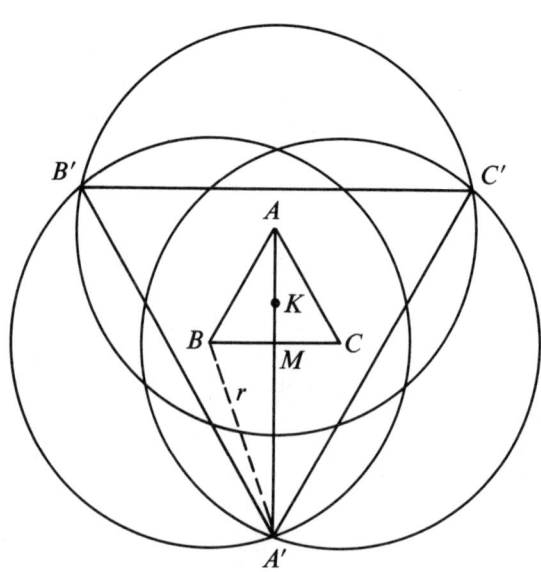

29. (E) By observing that $\left[x^3 + \dfrac{1}{x^3}\right]^2 = x^6 + 2 + \dfrac{1}{x^6}$, one sees that

$$f(x) = \frac{\left[\left(x + \dfrac{1}{x}\right)^3\right]^2 - \left[x^3 + \dfrac{1}{x^3}\right]^2}{\left(x + \dfrac{1}{x}\right)^3 + \left(x^3 + \dfrac{1}{x^3}\right)}$$

$$= \left(x + \frac{1}{x}\right)^3 - \left(x^3 + \frac{1}{x^3}\right) = 3\left(x + \frac{1}{x}\right).$$

Since

$$0 \leqslant \left(\sqrt{x} - \frac{1}{\sqrt{x}}\right)^2 = x + \frac{1}{x} - 2, \quad \text{we have} \quad 2 \leqslant x + \frac{1}{x},$$

and $f(x) = 3\left(x + \dfrac{1}{x}\right)$ has a minimum value of 6, which is taken on at $x = 1$.

30. (B) Let F be the point on the extension of side AB past B for which $AF = 1$. Since $AF = AC$ and $\angle FAC = 60°$, $\triangle ACF$ is equilateral. Let G be the point on line segment BF for which $\angle BCG = 20°$. Then $\triangle BCG$ is similar to $\triangle DCE$ and $BC = 2(EC)$. Also $\triangle FGC$ is congruent to $\triangle ABC$. Therefore,

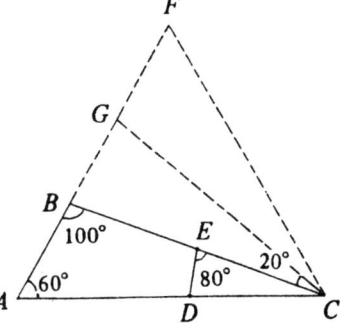

area $\triangle ACF = ($area $\triangle ABC +$ area $\triangle GCF) +$ area $\triangle BCG$,

$$\frac{\sqrt{3}}{4} = 2 \text{ area } \triangle ABC + 4 \text{ area } \triangle CDE,$$

$$\frac{\sqrt{3}}{8} = \text{area } \triangle ABC + 2 \text{ area } \triangle CDE.$$

1980 Solutions

1. (C) Since $\dfrac{100}{7} = 14 + \dfrac{2}{7}$, the number is 14.

2. (D) The highest powers of x in the factors $(x^2 + 1)^4$ and $(x^3 + 1)^3$ are 8 and 9, respectively. Hence the highest power of x in the product is $8 + 9 = 17$.

3. (E) If $\dfrac{2x - y}{x + y} = \dfrac{2}{3}$, then

$$6x - 3y = 2x + 2y,$$
$$4x = 5y,$$
$$\frac{x}{y} = \frac{5}{4}.$$

4. (C) The measures of angles $\angle ADC, \angle CDE$ and $\angle EDG$ are $90°$, $60°$ and $90°$, respectively. Hence the measure of $\angle GDA$ is

$$360° - (90° + 60° + 90°) = 120°.$$

5. (B) Triangle PQC is a $30°$–$60°$–$90°$ right triangle. Since $AQ = CQ$

$$\frac{PQ}{AQ} = \frac{PQ}{CQ} = \frac{1}{\sqrt{3}} = \frac{\sqrt{3}}{3}.$$

6. (A) Since x is positive the following are equivalent:

$$\sqrt{x} < 2x, \qquad x < 4x^2, \qquad 1 < 4x, \qquad \frac{1}{4} < x, \qquad x > \frac{1}{4}.$$

7 (B) By the Pythagorean theorem, diagonal AC has length 5. Since $5^2 + 12^2 = 13^2$, $\triangle DAC$ is a right triangle by the converse of the Pythagorean theorem. The area of $ABCD$ is

$$\left(\frac{1}{2}\right)(3)(4) + \left(\frac{1}{2}\right)(5)(12) = 36.$$

8. (A) The given equation implies each of these equations:

$$\frac{a + b}{ab} = \frac{1}{a + b},$$

$$(a + b)^2 = ab,$$

$$a^2 + ab + b^2 = 0.$$

Since the last equation is satisfied by the pairs (a, b) such that $a = \frac{1}{2}[-b \pm \sqrt{-3b^2}]$, and since the only real pair among these is $(0, 0)$, there are no pairs of real numbers satisfying the original equation.

9. (E) As shown in the adjoining figure, there are two possible starting points; therefore, x is not uniquely determined.

Possible starting points.

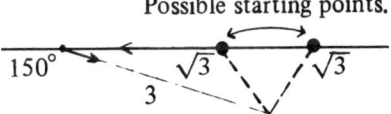

10. (D) Since the teeth are all the same size, equally spaced and are meshed, they all move with the same absolute speed v (v is the distance a point on the circumference moves per unit of time). Let α, β, γ be the angular speeds of A, B, C, respectively. If a, b, c represent the lengths of the circumferences of A, B, C, respectively, then

$$\alpha = \frac{v}{a}, \quad \beta = \frac{v}{b}, \quad \gamma = \frac{v}{c}.$$

Therefore, $\alpha a = \beta b = \gamma c$ or, equivalently,

$$\frac{\alpha}{\frac{1}{a}} = \frac{\beta}{\frac{1}{b}} = \frac{\gamma}{\frac{1}{c}}.$$

Thus the angular speeds are in the proportion

$$\frac{1}{a} : \frac{1}{b} : \frac{1}{c}.$$

Since a, b, c, are proportional to x, y, z, respectively, the angular speeds are in the proportion

$$\frac{1}{x} : \frac{1}{y} : \frac{1}{z}.$$

Multiplying each term by xyz yields the proportion

$$yz : xz : xy.$$

11. (D) The formula for the sum S_n of n terms of an arithmetic progression, whose first term is a and whose common difference is d, is $2S_n = n(2a + (n - 1)d)$. Therefore,

$$200 = 10(2a + 9d),$$
$$20 = 100(2a + 99d),$$
$$2S_{110} = 110(2a + 109d).$$

Subtracting the first equation from the second and dividing by 90 yields $2a + 109d = -2$. Hence $2S_{110} = 110(-2)$, so $S_{110} = -110$.

12. (C) In the adjoining figure, L_1 and L_2 intersect the line $x = 1$ at B and A, respectively; C is the intersection of the line $x = 1$ with the x-axis. Since $OC = 1$, AC is the slope of L_2 and BC is the slope of L_1. Therefore, $AC = n$, $BC = m$, and $AB = 3n$. Since OA is an angle bisector

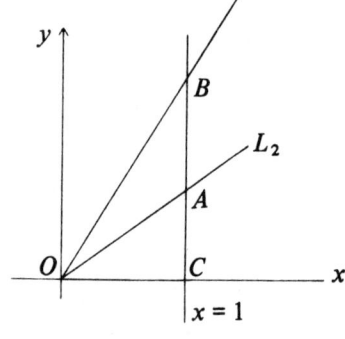

$$\frac{OC}{OB} = \frac{AC}{AB}.$$

This yields

$$\frac{1}{OB} = \frac{n}{3n} \quad \text{and} \quad OB = 3.$$

By the Pythagorean theorem $1 + (4n)^2 = 9$, so $n = \dfrac{\sqrt{2}}{2}$. Since $m = 4n$, $mn = 4n^2 = 2$.

OR

Let θ_1 and θ_2 be the angles of inclination of lines L_1 and L_2, respectively. Then $m = \tan\theta_1$ and $n = \tan\theta_2$. Since $\theta_1 = 2\theta_2$ and $m = 4n$,

$$4n = m = \tan\theta_1 = \tan 2\theta_2 = \frac{2\tan\theta_2}{1 - \tan^2\theta_2} = \frac{2n}{1 - n^2}.$$

Thus

$$4n = \frac{2n}{1 - n^2} \quad \text{and} \quad 4n(1 - n^2) = 2n.$$

Since $n \neq 0$, $2n^2 = 1$, and mn, which equals $4n^2$, is 2.

13. (B) If the bug travels indefinitely, the algebraic sum of the horizontal components of its moves approaches $\frac{4}{5}$, the limit of the geometric series

$$1 - \frac{1}{4} + \frac{1}{16} - \cdots = \frac{1}{1 - \left(-\frac{1}{4}\right)}.$$

Similarly, the algebraic sum of the vertical components of its moves approaches

$$\frac{2}{5} = \frac{1}{2} - \frac{1}{8} + \frac{1}{32} \cdots.$$

Therefore, the bug will get arbitrarily close to $\left(\frac{4}{5}, \frac{2}{5}\right)$.

OR

The line segments may be regarded as a complex geometric sequence with $a_1 = 1$ and $r = i/2$. Its sum is

$$\sum_{i=1}^{\infty} a_i = \frac{a_1}{1 - r} = \frac{2}{2 - i} = \frac{4 + 2i}{5} = \frac{4}{5} + \frac{2}{5}i.$$

In coordinate language, the limit is the point $\left(\frac{4}{5}, \frac{2}{5}\right)$.

Note: The figure shows that the limiting position (x, y) of the bug satisfies $x > 3/4$, $y > 3/8$. These inequalities alone prove that (B) is the correct choice.

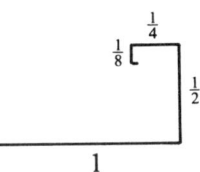

14. (A) For all $x \neq -\frac{3}{2}$,

$$x = f(f(x)) = \frac{c\left(\dfrac{cx}{2x + 3}\right)}{2\left(\dfrac{cx}{2x + 3}\right) + 3} = \frac{c^2 x}{2cx + 6x + 9},$$

which implies $(2c + 6)x + (9 - c^2) = 0$. Therefore, $2c + 6 = 0$ and $9 - c^2 = 0$. Thus, $c = -3$.

OR

The condition $f[f(x)] = x$ says that the function $f(x)$ is its own inverse. Thus, if

$$y = \frac{cx}{2x + 3}$$

is solved for x as a function of y, the result will be $x = f(y)$.

Indeed, we obtain

$$2xy + 3y - cx = 0, \qquad x(2y - c) = -3y,$$

and

$$x = \frac{-3y}{2y - c} = f(y),$$

which implies $c = -3$.

Note: In order to form $f[f(x)]$, we assumed that if $x \neq -3/2$, then $f(x) \neq -3/2$. This assumption can be justified *ex post facto* now that we have found $c = -3$; indeed if

$$f(x) = \frac{-3x}{2x + 3} = -\frac{3}{2},$$

then $6x = 6x + 9$, a contradiction.

15. (B) Let m be the price of the item in cents. Then $(1.04)m = 100n$. Thus $(8)(13)m = (100)^2 n$, so $m = (2)(5)^4 \frac{n}{13}$. Thus m is an integer if and only if 13 divides n.

16. (B) The edges of the tetrahedron are face diagonals of the cube. Therefore, if s is the length of an edge of the cube, the area of each face of the tetrahedron is

$$\frac{\left(s\sqrt{2}\,\right)^2 \sqrt{3}}{4} = \frac{s^2 \sqrt{3}}{2},$$

and the desired ratio is

$$\frac{6s^2}{4\left(\dfrac{s^2\sqrt{3}}{2}\right)} = \sqrt{3}.$$

17. (D) Since $i^2 = -1$,

$$(n + i)^4 = n^4 - 6n^2 + 1 + i(4n^3 - 4n).$$

This is real if and only if $4n^3 - 4n = 0$. Since $4n(n^2 - 1) = 0$ if and only if $n = 0, 1, -1$, there are only three values of n for which $(n + i)^4$ is real; $(n + i)^4$ is an integer in all three cases.

18. (D) $\log_b \sin x = a$; $\sin x = b^a$; $\sin^2 x = b^{2a}$; $\cos x = (1 - b^{2a})^{1/2}$;
$\log_b \cos x = \frac{1}{2} \log_b(1 - b^{2a})$.

19. (D) The adjoining figure is drawn and labelled according to the given data. We let r be the radius, x the distance from the center of the circle to the closest chord, and y the common distance between the chords. The Pythagorean theorem provides three equations in r, x, and y:

$$r^2 = x^2 + 10^2,$$
$$r^2 = (x + y)^2 + 8^2,$$
$$r^2 = (x + 2y)^2 + 4^2.$$

Subtracting the first equation from the second yields

$$0 = 2xy + y^2 - 36,$$

and subtracting the second equation from the third yields

$$0 = 2xy + 3y^2 - 48.$$

This last pair of equations yields $2y^2 - 12 = 0$, thus $y = \sqrt{6}$, and $x = 15/\sqrt{6}$. Finally

$$r = \sqrt{x^2 + 10^2} = 5\sqrt{22}/2.$$

20. (C) The number of ways of choosing 6 coins from 12 is

$$\binom{12}{6} = 924.$$

[The symbol $\binom{n}{k}$ denotes the number of ways k things may be selected from a set of n distinct objects.] "Having at least 50 cents" will occur if one of the following cases occurs:

    (1) Six dimes are drawn.
    (2) Five dimes and any other coin are drawn.
    (3) Four dimes and two nickels are drawn.

The numbers of ways (1), (2) and (3) can occur are $\binom{6}{6}$, $\binom{6}{5}\binom{6}{1}$ and $\binom{6}{4}\binom{4}{2}$, respectively. The desired probability is, therefore,

$$\frac{\binom{6}{6} + \binom{6}{4}\binom{4}{2} + \binom{6}{5}\binom{6}{1}}{924} = \frac{127}{924}.$$

21. (A) In the adjoining figure the line segment from E to G, the midpoint of DC, is drawn. Then

$$\text{area } \triangle EBG = \left(\frac{2}{3}\right)(\text{area } \triangle EBC),$$

$$\text{area } \triangle BDF = \left(\frac{1}{4}\right)(\text{area } \triangle EBG) = \left(\frac{1}{6}\right)(\text{area } \triangle EBC).$$

(Note that since EG connects the midpoints of sides AC and DC in $\triangle ACD$, EG is parallel to AD.) Therefore,

area $FDCE$

$$= \left(\frac{5}{6}\right)(\text{area } \triangle EBC)$$

and

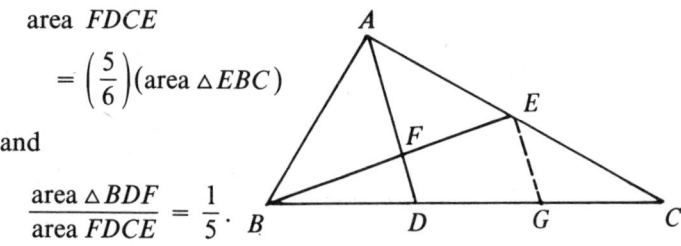

$$\frac{\text{area } \triangle BDF}{\text{area } FDCE} = \frac{1}{5}.$$

The measure of $\angle CBA$ was not needed.

22. (E) In the adjoining figure, the graphs of $y = 4x + 1$, $y = x + 2$ and $y = -2x + 4$ are drawn. The solid line represents the graph of the function f. Its maximum occurs at the intersection of the lines $y = x + 2$ and $y = -2x + 4$. Thus

$$x = \frac{2}{3}$$

and

$$f\left(\frac{2}{3}\right) = \frac{8}{3}.$$

23. (C) Applying the Pythagorean theorem to $\triangle CDF$ and $\triangle CEG$ in the adjoining figure yields

$$4a^2 + b^2 = \sin^2 x,$$
$$a^2 + 4b^2 = \cos^2 x.$$

Adding these equations, we obtain

$$5(a^2 + b^2) = \sin^2 x + \cos^2 x = 1.$$

Hence

$$AB = 3\sqrt{a^2 + b^2}$$

$$= 3\sqrt{\frac{1}{5}}$$

$$= 3\frac{\sqrt{5}}{5}.$$

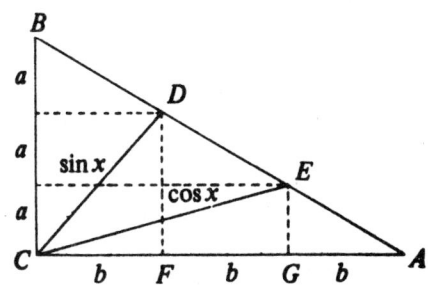

24. (D) Denote the given polynomial by $P(x)$. Since $P(x)$ is divisible by $(x - r)^2$, we have, for some polynomial $L(x)$,

$$P(x) = (x - r)^2 L(x),$$

and since $P(x)$ is of degree 3, $L(x)$ is of degree 1, say $L(x) = ax + b$. So we have the identity

$$8x^3 - 4x^2 - 42x + 45 = (x^2 - 2rx + r^2)(ax + b).$$

In the right member the coefficient of x^3 is a, and the constant term is br^2. Hence $a = 8$, and $br^2 = 45$, or $b = 45/r^2$, and the identity becomes

$$8x^3 - 4x^2 - 42x + 45 = (x^2 - 2rx + r^2)\left(8x + \frac{45}{r^2}\right).$$

Equating coefficients of x^2 and x, we obtain

$$-16r + \frac{45}{r^2} = -4, \qquad 8r^2 - \frac{90}{r} = -42.$$

Multiplying the first equation by $2r$ and adding it to the second, we have

$$-24r^2 = -8r - 42$$

or

$$12r^2 - 4r - 21 = 0 = (2r - 3)(6r + 7),$$

so that $r = 3/2$ or $r = -7/6$. Now $P(3/2) = 0$, but $P(-7/6) \neq 0$. We conclude that $r = 3/2 = 1.5$.

OR

For solvers familiar with calculus we include the following alternative. By hypothesis $8x^3 - 4x^2 - 42x + 45 = 8(x - r)^2(x - s)$. Differentiating both sides, we obtain

$$24x^2 - 8x - 42 = 16(x - r)(x - s) + 8(x - r)^2.$$

Setting $x = r$ and dividing by 2, we get $12r^2 - 4r - 21 = 0$. From here on the solution proceeds as above.

25. (C) The given information implies that $a_n > a_{n-1}$ if and only if $n + c$ is a perfect square. Since $a_2 > a_1$ and $a_5 > a_4$, it follows that $2 + c$ and $5 + c$ are both squares. The only squares differing by 3 are 1 and 4; hence $2 + c = 1$, so $c = -1$. Now

$$a_2 = 3 = b\left[\sqrt{2 + c}\right] + d = b\left[\sqrt{1}\right] + d = b + d.$$

Hence $b + c + d = 3 - 1 = 2$. (Although we only needed to find $b + d$ here, it is easy to see by setting $n = 1$ that $d = 1$, and hence $b = 2$.)

Note: The last member of the k-th string of equal terms occupies the position $1 + 3 + 5 + \cdots + 2k - 1 = k^2$ in the sequence. Its successor is $a_{k^2+1} = a_{k^2} + 2$. Therefore

$$\left[\sqrt{k^2 + 1 + c}\right] - \left[\sqrt{k^2 + c}\right] = \frac{2}{b} = 1,$$

so $c = -1$ and $b = 2$.

26. (E) A smaller regular tetrahedron circumscribing just one of the balls may be formed by introducing a plane parallel to the horizontal face as shown in Figure 1. Let the edge of this tetrahedron be t. Next construct the quadrilateral $C_1B_1B_2C_2$, where C_1 and C_2 are centers of two of the balls. By the symmetry of the ball tetrahedron configuration the sides C_1B_1 and C_2B_2 are parallel and equal. It follows that

$$B_1B_2 = C_1C_2 = 1 + 1 = 2.$$

Thus, $s = t + 2$, and our problem reduces to that of determining t.

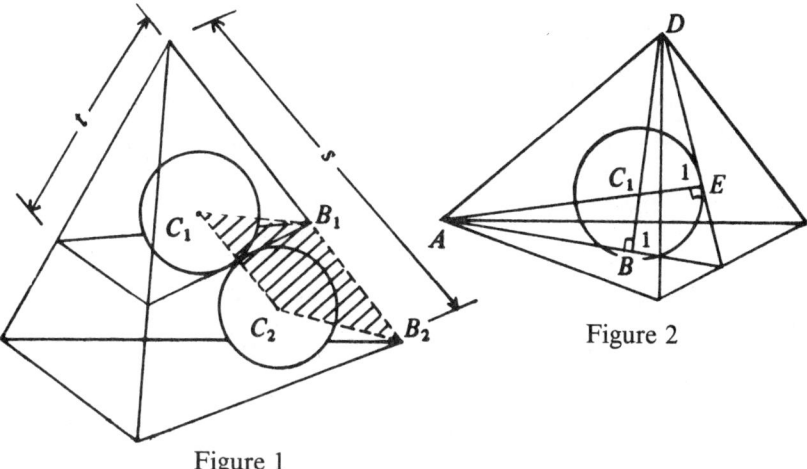

Figure 2

Figure 1

In $\triangle ABC_1$, in Figure 2, AC_1 has a length of one radius less than the length b of altitudes DB and AE of the smaller tetrahedron; and AB has length $\frac{2}{3}(\sqrt{3}/2)t$, since B is the center of a face. Applying the Pythagorean theorem to $\triangle ABC_1$ yields

$$(b - 1)^2 = 1^2 + \left[\frac{2}{3}\left(\frac{\sqrt{3}}{2}t\right)\right]^2,$$

and applying the Pythagorean theorem to $\triangle ADB$ yields

$$t^2 = b^2 + \left[\frac{2}{3}\left(\frac{\sqrt{3}}{2}t\right)\right]^2.$$

Solving the second equation for b, substituting this value for b in the first equation, and solving the resulting equation for the positive value of t yields $t = 2\sqrt{6}$. Thus $s = 2 + t = 2 + 2\sqrt{6}$.

OR

We first consider the regular tetrahedron T whose vertices are the centers of the four unit balls and calculate the distance d from the center of T to each face. We then magnify T so that the distance from the center to each face is increased by 1, the radius of the balls. We thus obtain the tetrahedron T' whose edge length e' is to be determined. Since T and T' are similar, and since the edges of T have length 2, we have

(1)
$$\frac{e'}{2} = \frac{d+1}{d}.$$

We carry out this plan by placing the center of T at the origin of a three dimensional coordinate system and denote the vectors from the origin to the vertices by c_1, c_2, c_3, c_4, and their common length by c. Since $c_1 + c_2 + c_3 + c_4 = 0$,

(2)
$$-c_1 = c_2 + c_3 + c_4.$$

Taking the scalar product of both sides with c_1 yields

$$-c_1 \cdot c_1 = -c^2 = c_1 \cdot c_2 + c_1 \cdot c_3 + c_1 \cdot c_4 = 3c^2\cos\theta,$$

where θ is the angle between c_i and c_j, $i \neq j$. Thus

$$\cos\theta = -\tfrac{1}{3}.$$

The square of the length of an edge of T is

$$|c_1 - c_2|^2 = 4 = 2c^2 - 2c^2\cos\theta = c^2\left(2 + \frac{2}{3}\right) = \frac{8}{3}c^2,$$

so

(3)
$$c = \sqrt{3/2}.$$

Now the line from the origin perpendicular to a face of T goes through the centroid of the triangular face. So the distance d is the length of the vector $\frac{1}{3}(c_2 + c_3 + c_4)$; by (2) we have,

$$d = \left|\tfrac{1}{3}(c_2 + c_3 + c_4)\right| = \tfrac{1}{3}c.$$

Substituting for c from (3), we obtain

$$d = (1/3)\sqrt{3/2} = 1/\sqrt{6}.$$

Finally, from (1),

$$e' = 2\frac{d+1}{d} = 2\frac{(1/\sqrt{6}) + 1}{1/\sqrt{6}} = 2 + 2\sqrt{6}.$$

OR

Some solvers may be familiar with the fact that the altitudes of a regular tetrahedron intersect at a point $3/4$ of the way from any vertex to the center of the opposite face.[†] Given this

† This can be proved by finding the center of mass M of four unit masses situated at the vertices. In finding M, the three masses on the bottom face can be replaced by a single mass of 3 units located at the center of the bottom face. Hence M is $3/4$ of the way from the top vertex to this latter point. The result now follows by symmetry.

fact, it follows that the altitude DB of the small tetrahedron is 4. From $\triangle ADB$ we find

$$t^2 = 4^2 + \left(\frac{2}{3}\frac{\sqrt{3}}{2}t\right)^2 = 16 + \frac{t^2}{3}.$$

Hence $\frac{2}{3}t^2 = 16$, or $t = \sqrt{24} = 2\sqrt{6}$.

27. (E) Let $a = \sqrt[3]{5 + 2\sqrt{13}}$, $b = \sqrt[3]{5 - 2\sqrt{13}}$ and $x = a + b$. Then

$$x^3 = a^3 + 3a^2b + 3ab^2 + b^3,$$
$$x^3 = a^3 + b^3 + 3ab(a + b),$$
$$x^3 = 10 + 3\sqrt[3]{-27}\,x.$$

The last equation is equivalent to $x^3 + 9x - 10 = 0$, or $(x - 1)(x^2 + x + 10) = 0$, whose only real solution is $x = 1$.

28. (C) Let $f(x) = x^2 + x + 1$ and $g_n(x) = x^{2n} + 1 + (x + 1)^{2n}$. Note that $(x - 1)f(x) = x^3 - 1$, so that the zeros of $f(x)$ are the complex cube roots of 1:

$$\omega = -\frac{1}{2} + \frac{\sqrt{3}}{2}i = \cos 120° + i\sin 120° = e^{i\pi/3}, \quad †$$

$$\omega' = -\frac{1}{2} - \frac{\sqrt{3}}{2}i = \cos 240° + i\sin 240° = e^{i2\pi/3} = \omega^2.$$

Note that

$$\omega^3 = (\omega')^3 = 1.$$

Now $g_n(x)$ is divisible by $f(x)$ if and only if $g_n(\omega) = g_n(\omega') = 0$. Since ω and ω' are complex conjugates, it suffices to determine those n for which

$$g_n(\omega) = \omega^{2n} + (\omega + 1)^{2n} + 1 = 0.$$

We note that

$$\omega + 1 = \frac{1}{2} + \frac{\sqrt{3}}{2}i = e^{i\pi/6}, \quad \text{so} \quad (\omega + 1)^2 = e^{i\pi/3} = \omega.$$

---

†The polar representation of the complex number $a + ib$ is $r(\cos\theta + i\sin\theta)$, where $r = \sqrt{a^2 + b^2}$ and $\theta = \arctan\frac{b}{a}$. For brevity, we set $\cos\theta + i\sin\theta = e^{i\theta}$, θ in radians. Note that $e^{i2k\pi} = 1$ for all integers k. This exponential notation is not arbitrary. Complex powers of e are defined as $e^{\alpha+i\theta} = e^{\alpha}(\cos\theta + i\sin\theta)$ for all real α and θ.

Thus, whenever n is a multiple of 3, say $n = 3k$, we have

$$g_{3k}(\omega) = \omega^{6k} + (\omega + 1)^{6k} + 1 = 1 + 1 + 1 = 3 \neq 0.$$

Suppose n is not a multiple of 3.
If $n = 3k + 1$,

$$\omega^{2n} = \omega^{6k+2} = \omega^2, \qquad (\omega + 1)^{2n} = \omega^n = \omega^{3k+1} = \omega,$$

and

$$g_{3k+1}(\omega) = \omega^2 + \omega + 1 = -1 + 1 = 0.$$

If $n = 3k + 2$,

$$\omega^{2n} = \omega^{6k+4} = \omega, \qquad (\omega + 1)^{2n} = \omega^{3k+2} = \omega^2,$$

and

$$g_{3k+2}(\omega) = \omega + \omega^2 + 1 = 0.$$

Thus $g_n(\omega) \neq 0$ if and only if n is a multiple of 3, and $g_n(x)$ fails to be divisible by $f(x)$ only in that case.

OR

Both of the given polynomials have integer coefficients, and $x^2 + x + 1$ has leading coefficient 1. Hence if

$$(1) \qquad x^{2n} + 1 + (x + 1)^{2n} = Q(x)(x^2 + x + 1),$$

then $Q(x)$ has integer coefficients. Setting $x = 2$ in (1), we obtain

$$2^{2n} + 1 + 3^{2n} = Q(2) \cdot 7,$$

which shows that 7 divides $2^{2n} + 1 + 3^{2n}$. However if n is divisible by 3, both 2^{2n} and 3^{2n} leave remainders of 1 upon division by 7, since $2^{6k} = 64^k = (7 \cdot 9 + 1)^k$, and $3^{6k} = (729)^k = (7 \cdot 104 + 1)^k$; so $2^{2n} + 1 + 3^{2n}$ leaves a remainder of 3 upon division by 7.

Note: The second solution, unlike the first, does not tell us what happens when n is not divisible by 3, but it does enable us to answer the question. We know that only one listed answer is correct, and (C) is a correct answer since 3 divides 21.

29. (A) If the last equation is multiplied by 3 and added to the first equation, we obtain

$$4x^2 + 2y^2 + 23z^2 = 331.$$

Clearly z^2 is odd and less than 25, so $z^2 = 1$ or 9. This leads to the two equations

$$2x^2 + y^2 = 154$$

and

$$2x^2 + y^2 = 62,$$

both of which are quickly found to have no solutions. Note that we made no use here of the second of the original equations.

OR

Adding the given equations and rearranging the terms of the resulting equation yields

$$\left(x^2 - 2xy + y^2\right) + \left(y^2 + 6yz + 9z^2\right) = 175$$

or

$$(x - y)^2 + (y + 3z)^2 = 175.$$

The square of an even integer is divisible by 4; the square of an odd integer, $(2n + 1)^2$, has remainder 1 when divided by 4. So the sum of two perfect squares can only have 0, 1 or 2 as a remainder when divided by 4. But 175 has remainder 3 upon division by 4, and hence the left and right sides of the equation above cannot be equal. Thus there are no integral solutions.

30. (B) That N is squarish may be expressed algebraically as follows: there are single digit integers A, B, C, a, b, c such that

$$N = 10^4A^2 + 10^2B^2 + C^2 = \left(10^2a + 10b + c\right)^2,$$

where each of A, B, C exceeds 3, and so a and c are positive. Since $10^2B^2 + C^2 < 10^4$ we can write

$$10^4A^2 < \left(10^2a + 10b + c\right)^2 < 10^4A^2 + 10^4 < 10^4(A + 1)^2.$$

Taking square roots we obtain

$$100A < 100a + 10b + c < 100A + 100,$$

from which it follows that $A = a$. Hence $a \geqslant 4$. Now consider

$$M = N - 10^4 A^2 = (10^2 a + 10b + c)^2 - 10^4 a^2$$
$$= 10^3 (2ab) + 10^2 (b^2 + 2ac) + 10(2bc) + c^2.$$

Since M has only four digits, $2ab < 10$, which implies that $ab \leqslant 4$. Thus either (i) $b = 0$, or (ii) $a = 4$ and $b = 1$. In case (ii),

$$N = (410 + c)^2 = 168100 + 820c + c^2.$$

If $c = 1$ or 2, the middle two digits of N form a number exceeding 81, hence not a square. If $c \geqslant 3$, then the leftmost two digits of N are 17. Therefore case (i) must hold, and we have

$$N = (10^2 a + c)^2 = 10^4 a^2 + 10^2 (2ac) + c^2.$$

Thus $a \geqslant 4$, $c \geqslant 4$ and $2ac$ is an even two-digit perfect square. It is now easy to check that either $a = 8$, $c = 4$, $N = 646416$, or $a = 4$, $c = 8$, $N = 166464$.

1981 Solutions

1. (E) $x + 2 = 4$; $(x + 2)^2 = 16$.

2. (C) The Pythagorean theorem, applied to $\triangle EBC$, yields $(BC)^2 = 2^2 - 1^2 = 3$. This is the area of the square.

3. (D) $\dfrac{1}{x} + \dfrac{1}{2x} + \dfrac{1}{3x} = \dfrac{6}{6x} + \dfrac{3}{6x} + \dfrac{2}{6x} = \dfrac{11}{6x}$.

4. (C) Let x be the larger number. Then $x - 8$ is the smaller number and $3x = 4(x - 8)$, so that $x = 32$.

5. (C) In $\triangle BDC, \angle BDC = 40°$. Since DC is parallel to AB,
$$\angle DBA = 40°.$$
Also, $\angle BAD = 40°$ since base angles of an isosceles triangle are equal. Therefore $\angle ADB = 100°$.

6. (A) $(y^2 + 2y - 2)x = (y^2 + 2y - 1)x - (y^2 + 2y - 1)$,
$$\left[(y^2 + 2y - 2) - (y^2 + 2y - 1)\right]x = -(y^2 + 2y - 1),$$
$$x = y^2 + 2y - 1.$$

OR

Rewrite the right member of the given equality as
$$\frac{(y^2 + 2y - 1)}{(y^2 + 2y - 1) - 1}$$
and note by inspection that $x = y^2 + 2y - 1$.

7. (B) The least common multiple of $2, 3, 4$ and 5 is 60. The numbers divisible by $2, 3, 4$ and 5 are integer multiples of 60.

8. (A) The given expression equals

$$\frac{1}{x+y+z}\left(\frac{1}{x}+\frac{1}{y}+\frac{1}{z}\right)\left(\frac{1}{xy+yz+zx}\right)\left(\frac{1}{xy}+\frac{1}{yz}+\frac{1}{zx}\right)$$

$$=\frac{1}{x+y+z}\left(\frac{xy+yz+zx}{xyz}\right)\left(\frac{1}{xy+yz+zx}\right)\left(\frac{x+y+z}{xyz}\right)$$

$$=\frac{1}{(xyz)^2}=x^{-2}y^{-2}z^{-2}.$$

Note: The given expression is homogeneous of degree -6; i.e. if x, y, z are multiplied by t, the expression is multiplied by t^{-6}. Only choice (A) has this property.

9. (A) Let s be the length of an edge of the cube, and let R and T be vertices of the cube as shown in the adjoining figure. Then applying the Pythagorean theorem to $\triangle PQR$ and $\triangle PRT$ yields

$$a^2-s^2=(PR)^2=s^2+s^2,$$

$$a^2=3s^2.$$

The surface area is $6s^2=2a^2$.

10. (E) If (p, q) is a point on line L, then by symmetry (q, p) must be a point on K. Therefore, the points on K satisfy

$$x=ay+b.$$

Solving for y yields

$$y=\frac{x}{a}-\frac{b}{a}.$$

11. (C) Let the sides of the triangle have lengths $s-d, s, s+d$. Then by the Pythagorean theorem

$$(s-d)^2+s^2=(s+d)^2.$$

Squaring and rearranging the terms yields

$$s(s-4d)=0.$$

Since s must be positive, $s=4d$. Thus the sides have lengths $3d, 4d, 5d$. Since the sides must have lengths divisible by 3, 4, or 5, only choice (C) could be the length of a side.

Note: The familiar 3–4–5 right triangle has sides which are in arithmetic progression. It is perhaps fairly natural to think of multiplying the sides by 27, thus getting 81, 108, 135. Since there is only one correct choice, it must be (C).

12. (E) The number obtained by increasing M by $p\%$ and decreasing the result by $q\%$ is $M\left(1 + \dfrac{p}{100}\right)\left(1 - \dfrac{q}{100}\right)$ and exceeds M if and only if the following equivalent inequalities hold:

$$M\left(1 + \frac{p}{100}\right)\left(1 - \frac{q}{100}\right) > M,$$

$$\left(1 + \frac{p}{100}\right)\left(1 - \frac{q}{100}\right) > 1,$$

$$1 + \frac{p}{100} > \frac{1}{1 - \dfrac{q}{100}} = \frac{100}{100 - q},$$

$$\frac{p}{100} > \frac{100}{100 - q} - 1 = \frac{q}{100 - q},$$

$$p > \frac{100q}{100 - q}.$$

13. (E) If A denotes the value of the unit of money at a given time, then $.9A$ denotes its value a year later and $(.9)^n A$ denotes its value n years later. We seek the smallest integer n such that n satisfies these equivalent inequalities:

$$(.9)^n A \leqslant .1A,$$

$$\left(\frac{9}{10}\right)^n \leqslant \frac{1}{10},$$

$$\log_{10}\left(\frac{9}{10}\right)^n \leqslant \log_{10}\frac{1}{10},$$

$$n(2\log_{10}3 - 1) \leqslant -1,$$

$$n \geqslant \frac{1}{1 - 2\log_{10}3} \approx 21.7.$$

14. (A) Let a and r be the first term and the common ratio of successive terms in the geometric sequence, respectively. Then

$$a + ar = 7,$$

$$\frac{a(r^6 - 1)}{r - 1} = 91.$$

The result of dividing the second equation by the first is

$$\frac{a(r^6 - 1)}{(r - 1)a(r + 1)} = \frac{91}{7} = 13.$$

The left member reduces to

$$\frac{(r^2 - 1)(r^4 + r^2 + 1)}{r^2 - 1} = r^4 + r^2 + 1,$$

so

$$r^4 + r^2 - 12 = 0,$$
$$(r^2 + 4)(r^2 - 3) = 0.$$

Thus $r^2 = 3$ and

$$a + ar + ar^2 + ar^3 = (a + ar)(1 + r^2)$$
$$= 7(4) = 28.$$

15. (B) For this solution write log for $\log_b$. The given equation is equivalent to

$$(2x)^{\log 2} = (3x)^{\log 3}.$$

Equating the logarithms of the two sides, one obtains

$$\log 2(\log 2x) = \log 3(\log 3x),$$

$$\log 2(\log 2 + \log x) = \log 3(\log 3 + \log x),$$

$$(\log 2)^2 + \log 2(\log x) = (\log 3)^2 + \log 3(\log x),$$

$$(\log 2)^2 - (\log 3)^2 = (\log 3 - \log 2)\log x,$$

$$-(\log 2 + \log 3) = \log x,$$

$$\log \frac{1}{6} = \log x,$$

$$\frac{1}{6} = x.$$

16. (E) Grouping the base three digits of x in pairs yields

$$x = (1 \cdot 3^{19} + 2 \cdot 3^{18}) + (1 \cdot 3^{17} + 1 \cdot 3^{16})$$
$$+ \cdots + (2 \cdot 3 + 2)$$
$$= (1 \cdot 3 + 2)(3^2)^9 + (1 \cdot 3 + 1)(3^2)^8$$
$$+ \cdots + (2 \cdot 3 + 2).$$

Therefore, the first base nine digit of x is $1 \cdot 3 + 2 = 5$.

17. (B) Replacing x by $\dfrac{1}{x}$ in the given equation

$$f(x) + 2f\left(\frac{1}{x}\right) = 3x$$

yields

$$f\left(\frac{1}{x}\right) + 2f(x) = \frac{3}{x}.$$

Eliminating $f\left(\dfrac{1}{x}\right)$ from the two equations yields

$$f(x) = \frac{2 - x^2}{x}.$$

Then $f(x) = f(-x)$ if and only if

$$\frac{2 - x^2}{x} = \frac{2 - (-x)^2}{-x},$$

or $x^2 = 2$. Thus $x = \pm \sqrt{2}$ are the only solutions.

18. (C) We have $\dfrac{x}{100} = \sin x$ if and only if $\dfrac{-x}{100} = \sin(-x)$; thus, the given equation has an equal number of positive and negative solutions. Also $x = 0$ is a solution. Furthermore, all positive solutions are less than or equal to 100, since

$$|x| = 100|\sin x| \leqslant 100.$$

Since $15.5 \leqslant 100/(2\pi) \leqslant 16$, the graphs of $x/100$ and $\sin x$ are as shown in the figure on p. 158. Thus there is one solution to the given equation between 0 and π and two solutions in each of the intervals from $(2k - 1)\pi$ to $(2k + 1)\pi, 1 \leqslant k \leqslant 15$. The total number of solutions is, therefore,

$$1 + 2(1 + 2 \cdot 15) = 63.$$

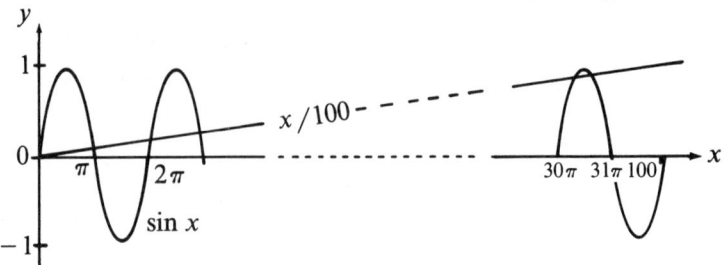

19. (B) In the adjoining figure, BN is extended
past N and meets AC
at E. Triangle BNA is
congruent to $\triangle ENA$,
since $\angle BAN = \angle EAN$,

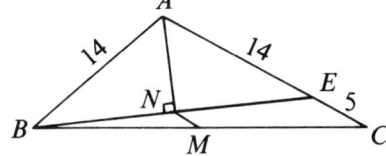

$AN = AN$ and $\angle ANB = \angle ANE$. Therefore N is the midpoint of BE, and $AB = AE = 14$. Thus $EC = 5$. Since
MN is the line joining the midpoints of sides BC and BE of
$\triangle CBE$, its length is $\frac{1}{2}(EC) = \frac{5}{2}$.

20. (B) Let $\angle DAR_1 = \theta$ and let θ_i be the (acute) angle the light
beam and the reflecting line form at the ith point of reflection. Applying the theorem on exterior angles of triangles
to $\triangle AR_1D$, then successively to the triangles $R_{i-1}R_iD$,
$2 \leqslant i \leqslant n$, and finally to $\triangle R_nBD$ yields

$$\theta_1 = \theta + 8°,$$
$$\theta_2 = \theta_1 + 8° = \theta + 16°,$$
$$\theta_3 = \theta_2 + 8° = \theta + 24°,$$
$$- - - - - - - - - - - - - - - -$$
$$\theta_n = \theta_{n-1} + 8° = \theta + (8n)°,$$
$$90° = \theta_n + 8° = \theta + (8n + 8)°.$$

But θ must be positive. Therefore,

$$0 \leqslant \theta = 90° - (8n + 8)°,$$
$$n \leqslant \frac{82}{8} < 11.$$

The maximum value of n, 10, occurs when $\theta = 2°$.

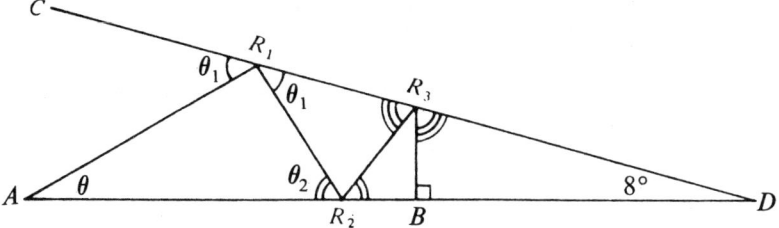

21. (D) Let θ be the angle opposite the side of length c. Now

$$(a + b + c)(a + b - c) = 3ab,$$

$$(a + b)^2 - c^2 = 3ab,$$

$$a^2 + b^2 - ab = c^2.$$

But

$$a^2 + b^2 - 2ab\cos\theta = c^2,$$

so that $ab = 2ab\cos\theta$, $\cos\theta = \dfrac{1}{2}$, and $\theta = 60°$.

22. (D) Consider the smallest cube containing all the lattice points (i, j, k), $1 \leqslant i, j, k \leqslant 4$, in a three dimensional Cartesian coordinate system. There are 4 main diagonals. There are 24 diagonal lines parallel to a coordinate plane: 2 in each of four planes parallel to each of the three coordinate planes. There are 48 lines parallel to a coordinate axis: 16 in each of the three directions. Therefore, there are $4 + 24 + 48 = 76$ lines.

OR

Let S be the set of lattice points (i, j, k) with $1 \leqslant i, j, k \leqslant 4$, and let T be the set of lattice points (i, j, k) with $0 \leqslant i, j, k \leqslant 5$. Every line segment containing four points of S can be extended at both ends so as to contain six points of T.

· Points of S
∘ Points of $T - S$

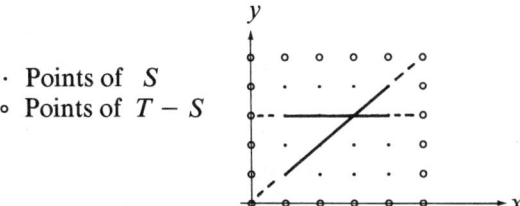

(The figure on p. 159 shows two of these extended lines in the xy-plane.) Every point of the "border" $T - S$ is contained in exactly one of these lines. Hence the number of lines is half the number of points in the border, namely half the number of points in T minus half the number of points in S

$$= \frac{1}{2}(6^3 - 4^3) = 76.$$

Note: The same reasoning shows that the number of ways of making tic-tac-toe on an n-dimensional "board" of s^n lattice points is $\frac{1}{2}[(s + 2)^n - s^n]$.

23. (C) Let O and H be the points at which PQ and BC, respectively, intersect diameter AT. Sides AB and AC form a portion of the equilateral triangle circumscribing the smaller circle and tangent to the smaller circle at T. Therefore, $\triangle PQT$ is an equilateral triangle. Since $\triangle APQ$ is an equilateral triangle with a side in common with $\triangle PQT$, $\triangle APQ \cong \triangle PQT$. Thus $AO = OT$ and O is the center of the larger circle. This implies

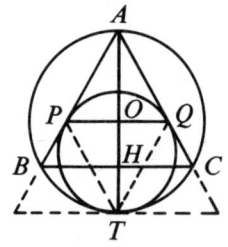

$$AO = \frac{2}{3}(AH), \quad \text{so that} \quad PQ = \frac{2}{3}(BC) = 8.$$

24. (D) Write $x + \dfrac{1}{x} = 2\cos\theta$ as $x^2 - 2x\cos\theta + 1 = 0$. Then

$$x = \cos\theta \pm \sqrt{\cos^2\theta - 1} = \cos\theta \pm i\sin\theta \,(= e^{\pm i\theta}).$$

By De Moivre's theorem

$$x^n = \cos n\theta \pm i\sin n\theta (= e^{\pm in\theta}),$$

$$\frac{1}{x^n} = \frac{1}{\cos n\theta \pm i\sin n\theta} = \cos n\theta \mp i\sin n\theta (= e^{\mp i\theta}).$$

Thus

$$x^n + \frac{1}{x^n} = 2\cos n\theta.$$

Note 1: Squaring both sides of the given equation, we obtain

$$x^2 + 2 + \frac{1}{x^2} = 4\cos^2\theta.$$

Hence

$$x^2 + \frac{1}{x^2} = 4\cos^2\theta - 2 = 2(2\cos^2\theta - 1) = 2\cos 2\theta.$$

This eliminates all choices except (D).

Note 2: Several contestants wrote to the Contests Committee to say that this problem had no solution because, for $0 < \theta < \pi$ as given, $2\cos\theta < 2$, yet $x + \frac{1}{x} \geqslant 2$ for all x. This claim is true for *real* x, but nowhere in the problem did it say that x must be real. When such a restriction is intended, it is always stated.

25. (A) In the adjoining figure let $\angle BAC = 3\alpha$, $c = AB$, $y = AD$, $z = AE$ and $b = AC$. Then by the angle bisector theorem[†]

(1) $\dfrac{c}{z} = \dfrac{2}{3}$ and $\dfrac{y}{b} = \dfrac{1}{2}$;

so

$z = \dfrac{3}{2}c$, $b = 2y$.

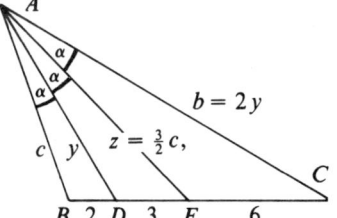

$$b = 2y$$
$$z = \tfrac{3}{2}c,$$

Using the law of cosines in $\triangle ADB$, $\triangle AED$ and $\triangle ACE$, respectively, yields the following expressions for $\cos\alpha$:

$$\frac{c^2 + y^2 - 4}{2cy} = \frac{\frac{9}{4}c^2 + y^2 - 9}{3cy} = \frac{\frac{9}{4}c^2 + 4y^2 - 36}{6cy}.$$

The equality of the first and second expressions implies

$$3c^2 - 2y^2 = 12.$$

[†] The angle bisector theorem states that each angle bisector of a triangle divides the opposite side into segments proportional in length to the adjacent sides, i.e.

(*) $\dfrac{d}{c} = \dfrac{e}{b}$

in the adjoining figure. This is a consequence of the law of sines applied to each subtriangle and the fact that the supplementary angles at D have equal sines.

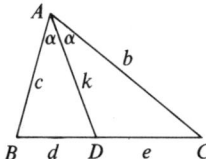

The equality of the first and third expressions implies

$$3c^2 - 4y^2 = -96.$$

Solving these two equations for c^2 and y^2 yields

$$c^2 = 40, \quad y^2 = 54.$$

Thus the sides of the triangle are

$$AB = c = 2\sqrt{10} \approx 6.3,$$
$$AC = b = 2y = 2\sqrt{54} = 6\sqrt{6} \approx 14.7,$$
$$BC = 11.$$

OR

Using the angle bisector formula[†] and (1) above, we have

$$y^2 + 6 = cz = \frac{2}{3}z^2, \qquad z^2 + 18 = yb = 2y^2.$$

Solving these equations for y^2 and z^2 yields $y^2 = 54$, $z^2 = 90$. Therefore

$$AB = c = \frac{2}{3}\sqrt{90} = 2\sqrt{10}, \quad AC = b = 2\sqrt{54} = 6\sqrt{6}.$$

[†] The angle bisector formula states that the square of the angle bisector plus the product of the segments of the opposite side is equal to the product of the adjacent sides:

(**) $$k^2 + de = bc.$$

This is a consequence of the angle bisector theorem and the law of cosines:

$$c^2 + k^2 - 2ck\cos\alpha = d^2,$$
$$b^2 + k^2 - 2bk\cos\alpha = e^2.$$

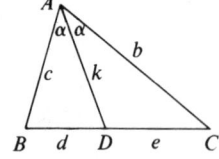

Multiplying the first equation by b, the second by c and subtracting, we obtain

$$bc^2 - b^2c - k^2(c - b) = d^2b - e^2c.$$

By (*) (see footnote on p. 161) the right member can be written $dec - edb$; thus

$$(c - b)[bc - k^2] = (c - b)de.$$

Dividing by $(c - b)$ and adding k^2 yields formula (**). In the isosceles case, $c - b = 0$; but the angle bisector is the altitude so that (**) becomes a consequence of the Pythagorean theorem.

26. (D) The probability that the first 6 is tossed on the k-th toss is the product

$$\begin{pmatrix} \text{probability that never a 6 was} \\ \text{tossed in the previous } k-1 \text{ tosses} \end{pmatrix} \begin{pmatrix} \text{probability that a 6 is} \\ \text{tossed on the } k^{\text{th}} \text{ toss} \end{pmatrix}$$

$$= (5/6)^{k-1} (1/6).$$

The probability that Carol will toss the first 6 is the sum of the probabilities that she will toss the first 6 on her first turn (3rd toss of the game), on her second turn (6th toss of the game), on her third turn, etc. This sum is

$$\left(\frac{5}{6}\right)^2 \frac{1}{6} + \left(\frac{5}{6}\right)^5 \frac{1}{6} + \cdots + \left(\frac{5}{6}\right)^{3n-1} \frac{1}{6} + \cdots,$$

an infinite geometric series with first term $a = \left(\frac{5}{6}\right)^2 \frac{1}{6}$ and common ratio $r = \left(\frac{5}{6}\right)^3$. This sum is

$$\frac{a}{1-r} = \frac{5^2/6^3}{1-(5^3/6^3)} = \frac{5^2}{6^3-5^3} = \frac{25}{91}.$$

27. (C) In the figure, line segment DC is drawn. Since $\overset{\frown}{AC} = 150°$, $\overset{\frown}{AD} = \overset{\frown}{AC} - \overset{\frown}{DC} = 150° - 30° = 120°$. Hence $\angle ACD = 60°$. Since $AC = DG$, $\overset{\frown}{GA} = \overset{\frown}{GD} - \overset{\frown}{AD} = \overset{\frown}{AC} - 120° = 30°$. Therefore $\overset{\frown}{CG} = 180°$ and $\angle CDG = 90°$. So $\triangle DEC$ is a 30°–60°–90° triangle.

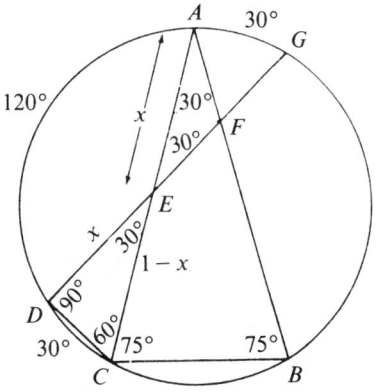

Since we are looking for the ratio of the areas, let us assume without loss of generality that $AC = AB = DG = 1$. Since AC and DG are chords of equal length in a circle, we have $AE = DE$. Let x be their common length. Then $CE = 1 - x = 2x/\sqrt{3}$. Solving for x yields $AE = x = 2\sqrt{3} - 3$. Since BA and DG are chords of equal length in a circle, we have $FG = FA$, and since $\triangle FAE$ is isosceles, $EF = FA$. Thus

$$EF = FG = \tfrac{1}{2}(1 - x).$$

Therefore

$$\text{area } \triangle AFE = \frac{1}{2}(AE)(AF)\sin 30°$$

$$= \frac{1}{2}x\frac{1-x}{2}\cdot\frac{1}{2} = \frac{x-x^2}{8} = \frac{7\sqrt{3}-12}{4};$$

$$\text{area } \triangle ABC = \frac{1}{2}(AB)(AC)\sin 30° = \frac{1}{4}.$$

Hence

$$\frac{\text{area } \triangle AFE}{\text{area } \triangle ABC} = 7\sqrt{3} - 12.$$

28. (D) Let $g(x) = x^3 + a_2x^2 + a_1x + a_0$ be an arbitrary cubic with constants of the specified form. Because x^3 dominates the other terms for large enough x, $g(x) > 0$ for all x greater than the largest real root of g. Thus we seek a particular g in which the terms $a_2x^2 + a_1x + a_0$ "hold down" $g(x)$ as much as possible, so that the value of the largest real root is as large as possible. This suggests that the answer to the problem is the largest root of $f(x) = x^3 - 2x^2 - 2x - 2$. Call this root r_0. Since $f(0) = -2$, r_0 is certainly positive. To verify this conjecture, note that for $x \geqslant 0$,

$$-2x^2 \leqslant a_2x^2, \qquad -2x \leqslant a_1x, \quad \text{and} \quad -2 \leqslant a_0.$$

Summing these inequalities and adding x^3 to both sides yields $f(x) \leqslant g(x)$ for all $x \geqslant 0$. Thus for all $x > r_0$, $0 < f(x) \leqslant g(x)$. That is, no g has a root larger than r_0, so r_0 is the r of the problem.

A sketch of f shows that it has a typical cubic shape, with largest root a little less than 3. In fact, $f(2) = -6$ and $f(3) = 1$. To be absolutely sure the answer is (D), not (C), compute $f(\frac{5}{2})$ to see if it is negative. Indeed,

$$f\left(\frac{5}{2}\right) = -\frac{31}{8}.$$

29. (E) Since x is the principal square root of some quantity, $x \geqslant 0$. For $x \geqslant 0$, the given equation is equivalent to

$$a - \sqrt{a + x} = x^2$$

or

$$a = \sqrt{a + x} + x^2.$$

The left member is a constant, the right member is an

increasing function of x, and hence the equation has exactly one solution. We write

$$\sqrt{a + x} = a - x^2,$$

$$\sqrt{a + x} + x = (a + x) - x^2,$$

$$= (\sqrt{a + x} + x)(\sqrt{a + x} - x).$$

Since $\sqrt{a + x} + x > 0$, we may divide by it to obtain

$$1 = \sqrt{a + x} - x \quad \text{or} \quad x + 1 = \sqrt{a + x},$$

so

$$x^2 + 2x + 1 = a + x,$$

and

$$x^2 + x + 1 - a = 0.$$

Therefore $x = \dfrac{-1 \pm \sqrt{4a - 3}}{2}$, and the positive root is x

$= \dfrac{-1 + \sqrt{4a - 3}}{2}$, the only solution of the original equation. Therefore, this is also the sum of the real solutions.

OR

As above, we derive $a - \sqrt{a + x} = x^2$, and hence $a - x^2 = \sqrt{a + x}$. Squaring both sides, we find that

$$a^2 - 2x^2 a + x^4 = a + x.$$

This is a quartic equation in x, and therefore not easy to solve; but it is only quadratic in a, namely

$$a^2 - (2x^2 + 1)a + x^4 - x = 0.$$

Solving this by the quadratic formula, we find that

$$a = \tfrac{1}{2}\left[2x^2 + 1 + \sqrt{4x^4 + 4x^2 + 1 - 4x^4 + 4x}\right]$$

$$= x^2 + x + 1.$$

[We took the positive square root since $a > x^2$; indeed $a - x^2 = \sqrt{a + x}$.]

Now we have a quadratic equation for x, namely

$$x^2 + x + 1 - a = 0,$$

which we solve as in the previous solution.

Note: One might notice that when $a = 3$, the solution of the original equation is $x = 1$. This eliminates all choices except (E).

30. (D) Since the coefficient of x^3 in the polynomial function $f(x) = x^4 - bx - 3$ is zero, the sum of the roots of $f(x)$ is zero, and therefore

$$\frac{a+b+c}{d^2} = \frac{a+b+c+d-d}{d^2} = \frac{-1}{d}.$$

Similarly,

$$\frac{a+c+d}{b^2} = \frac{-1}{b}, \quad \frac{a+b+d}{c^2} = \frac{-1}{c}, \quad \frac{b+c+d}{a^2} = \frac{-1}{a}.$$

Hence the equation $f\left(-\dfrac{1}{x}\right) = 0$ has the specified solutions:

$$\frac{1}{x^4} + \frac{b}{x} - 3 = 0,$$

$$1 + bx^3 - 3x^4 = 0,$$

$$3x^4 - bx^3 - 1 = 0.$$

1982 Solutions

1. (E)

$$x^2 + 0x - 2 \overline{\smash{\big)}\, x^3 + 0x^2 + 0x - 2}$$

$$\begin{array}{r} x \\ x^2 + 0x - 2 \enclose{longdiv}{x^3 + 0x^2 + 0x - 2} \\ \underline{x^3 + 0x^2 - 2x} \\ 2x - 2 = \text{remainder.} \end{array}$$

OR

We have

$$\frac{x^3 - 2}{x^2 - 2} = \frac{x^3 - 2x + 2x - 2}{x^2 - 2} = x + \frac{2x - 2}{x^2 - 2},$$

so the remainder is $2x - 2$.

2. (A) The answer is $\dfrac{8x + 2}{4} = 2x + \dfrac{1}{2}$.

3. (C) For $x = 2$, the expression equals

$$(2^2)^{(2^2)} = 4^4 = 4 \cdot 4 \cdot 4 \cdot 4 = 256.$$

4. (E) Let r be the radius of the semicircle. The perimeter of a semicircular region is $\pi r + 2r$. The area of the region is $\frac{1}{2}\pi r^2$. Therefore

$$\pi r + 2r = \frac{\pi r^2}{2},$$

$$2\pi + 4 = \pi r,$$

$$2 + \frac{4}{\pi} = r.$$

5. (C) Since $y = \dfrac{b}{a}x$, $\dfrac{b}{a} > 1$, and $x > 0$, it follows that x is the smaller number. Also, $x + y = c$. Thus

$$x + \frac{b}{a}x = c,$$

$$ax + bx = ac,$$

$$x = \frac{ac}{a + b}.$$

6. **(D)** The sum of the angles in a convex polygon of n sides is $(n - 2)180°$. Therefore, if x is the unknown angle,

$$(n - 2)180° = 2570° + x, \quad \text{with } 0° < x < 180°.$$

If $n = 17$, then $(n - 2)180° = 15(180°) = 2700°$ and $x = 130°$. Smaller values of n would yield negative values of x, and larger values of n would yield values of x greater than $180°$.

7. **(B)**
$$x * (y + z) = (x + 1)(y + z + 1) - 1,$$
$$(x * y) + (x * z) = [(x + 1)(y + 1) - 1]$$
$$+ [(x + 1)(z + 1) - 1]$$
$$= (x + 1)(y + z + 2) - 2.$$

Therefore, $x * (y + z) \neq (x * y) + (x * z)$. The remaining choices can easily be shown to be true.

8. **(B)** Since

$$\binom{n}{2} - \binom{n}{1} = \binom{n}{3} - \binom{n}{2},$$

we have

$$\frac{n(n - 1)}{2} - n = \frac{n(n - 1)(n - 2)}{6} - \frac{n(n - 1)}{2}.$$

Thus

$$n^3 - 9n^2 + 14n = 0,$$
$$n(n - 2)(n - 7) = 0.$$

Since $n > 3$, $n = 7$ is the solution.

(The answer may also be obtained by evaluating the sequence $\binom{n}{1}, \binom{n}{2}, \binom{n}{3}$ for the values of n listed as choices.)

9. **(B)** In the adjoining figure, ABC is the given triangle and $x = a$ is the dividing line. Since area $\triangle ABC = \frac{1}{2}(1)(8) = 4$, the two regions must each have area 2. Since the portion of

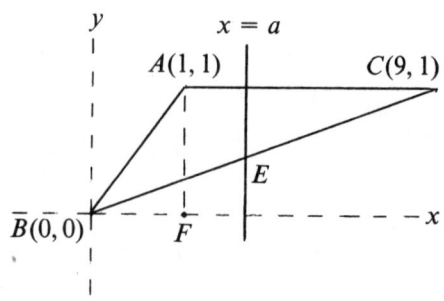

$\triangle ABC$ to the left of the vertical line through vertex A has area less than area $\triangle ABF = \frac{1}{2}$, the line $x = a$ is indeed right of A as shown. Since the equation of line BC is $y = x/9$, the vertical line $x = a$ intersects BC at a point $E:(a, a/9)$. Thus

$$\text{area } \triangle DEC = 2 = \frac{1}{2}\left(1 - \frac{a}{9}\right)(9 - a),$$

or

$$(9 - a)^2 = 36.$$

Then $9 - a = \pm 6$, and $a = 15$ or 3. Since the line $x = a$ must intersect $\triangle ABC$, $x = 3$.

10. (A) Since MN is parallel to BC,

$$\angle MOB = \angle CBO = \angle OBM,$$

and

$$\angle CON = \angle OCB = \angle NCO.$$

Therefore $MB = MO$ and $ON = NC$. Hence

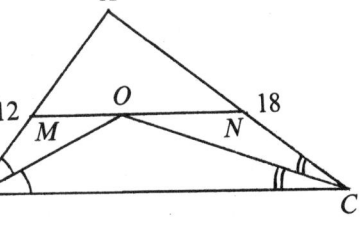

$$AM + MO + ON + AN = (AM + MB) + (AN + NC)$$

$$= AB + AC = 12 + 18 = 30.$$

Note that the given value $BC = 24$ was not needed; in fact, only the *sum* of the lengths of the other two sides was used.

11. (C) From the set $\{0, 1, \ldots, 9\}$ there are sixteen pairs of numbers $\{(0, 2), (2, 0), (1, 3), (3, 1), \ldots\}$ whose difference is ± 2. All but $(0, 2)$ can be used as the first and last digit, respectively, of the required number. For each of the 15 ordered pairs there are $8 \cdot 7 = 56$ ways to fill the remaining middle two digits. Thus there are $15 \cdot 56 = 840$ numbers of the required form.

12. (A) Since $f(x) = ax^7 + bx^3 + cx - 5$,

$$f(-x) = a(-x)^7 + b(-x)^3 + c(-x) - 5.$$

Therefore, $f(x) + f(-x) = -10$ and $f(7) + f(-7) = -10$. Hence, since $f(-7) = 7$, $f(7) = -17$.

13. (D) We have

$$p(\log_b a) = \log_b(\log_b a),$$
$$\log_b(a^p) = \log_b(\log_b a),$$
$$a^p = \log_b a.$$

14. (C) In the adjoining figure, MN is perpendicular to AG at M, and NF and PG are radii. Since $\triangle AMN \sim \triangle AGP$, it follows that $\dfrac{MN}{AN} = \dfrac{GP}{AP}$, or $\dfrac{MN}{45} = \dfrac{15}{75}$. Thus $MN = 9$. Applying the Pythagorean theorem to triangle MNF yields $(MF)^2 = (15)^2 - 9^2 = 144$, so $MF = 12$. Therefore $EF = 24$.

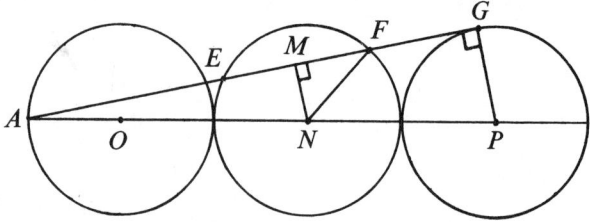

15. (D) We have

$$2[x] + 3 = 3[x - 2] + 5,$$
$$2[x] + 3 = 3([x] - 2) + 5,$$
$$[x] = 4.$$

Therefore, $4 < x < 5$, and $y = 2[x] + 3 = 11$. Hence, $15 < x + y < 16$. (Alternatively, if one draws the graphs of $y = 2[x] + 3$ and $y = 3[x - 2] + 5$, one can see that they overlap when $4 < x < 5$).

16. (B) Each exterior unit square which is removed exposes 4 interior unit squares, so the entire surface area in square meters is

$$6 \cdot 3^2 - 6 + 24 = 72.$$

17. (C) Let $y = 3^x$; then $3^{2x+2} = 9y^2$ and $3^{x+3} = 27y$. The given equation now becomes

$$9y^2 - 28y + 3 = (9y - 1)(y - 3) = 0$$

and has the solutions $y = 3$ and $y = \frac{1}{9}$. Hence the original equation has exactly two solutions, namely $x = 1$ and $x = -2$.

18. (D) Without loss of general- ity, let $HF = 1$ in the adjoining figure. Then $BH = 2$, $BF = \sqrt{3} = DG = GH$, and $DH = \sqrt{6}$. Since $DC = HC = \sqrt{3}$, $\triangle DCB \cong \triangle HCB$ and $DB = HB = 2$. Since $\triangle DBH$ is isosceles,

$$\cos\theta = \frac{\frac{1}{2}\sqrt{6}}{2} = \frac{\sqrt{6}}{4}.$$

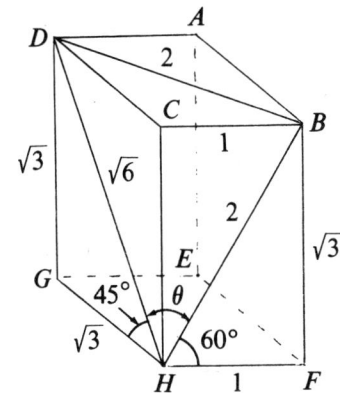

19. (B) When $2 \leqslant x \leqslant 3$, $f(x) = (x - 2) - (x - 4) + (2x - 6) = -4 + 2x$. Similar algebra shows that when $3 \leqslant x \leqslant 4$, $f(x) = 8 - 2x$; and when $4 \leqslant x \leqslant 8$, $f(x) = 0$. The graph of $f(x)$ in the adjoin- ing figure shows that the maximum and minimum of $f(x)$ are 2 and 0, respectively.

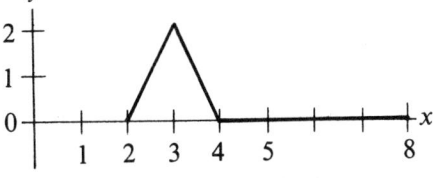

Note: Since a linear function reaches its extreme values at the endpoints of an interval, and since the given function is linear in each subin- terval, it suffices to calculate $f(2) = 0$, $f(3) = 2$, $f(4) = f(8) = 0$.

20. (D) Since

$$x^2 + y^2 = x^3,$$
$$y^2 = x^2(x - 1).$$

Therefore, if k is an integer satisfying $x - 1 = k^2$, i.e., $x = 1 + k^2$, then there is a y satisfying $x^2 + y^2 = x^3$. Hence there are infinitely many solutions.

21. (E) Since the medians of a triangle intersect at a point two thirds the distance from the vertex and one third the distance from the side to which they are drawn, we can let $x = DN$ and

$2x = BD$. Right triangles BCN and BDC are similar, so $2x = BD$. Right triangles BCN and BDC are similar, so

$$\frac{s}{3x} = \frac{2x}{s}.$$

Thus $s^2 = 6x^2$, or $x = \dfrac{s}{\sqrt{6}}$,

and $BN = 3x = \dfrac{3s}{\sqrt{6}} = \dfrac{s\sqrt{6}}{2}$.

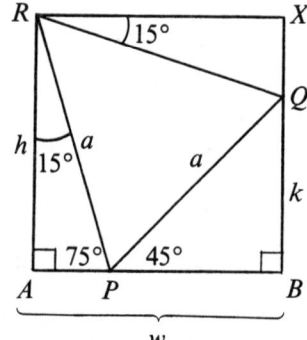

22. (E) In the adjoining figure RX is perpendicular to QB at X. Since $\angle QPR = 60°$, $\triangle RPQ$ is equilateral and $RQ = a$. Also $\angle ARP = \angle QRX = 15°$. Therefore, $\triangle RXQ \cong \triangle RAP$. Thus $w = h$.

OR

$w = AP + PB = a\cos 75° + a\cos 45°$.
From the identity

$$\cos A + \cos B = 2\cos\frac{A+B}{2}\cos\frac{A-B}{2},$$

it follows that

$$w = a\,2\cos\frac{75° + 45°}{2}\cos\frac{75° - 45°}{2}$$

$$= a\,2\cos 60°\cos 15° = a\cos 15° = a\sin 75° = h.$$

23. (A) In the adjoining figure, n denotes the length of the shortest side, and θ denotes the measure of the smallest angle. Using the law of sines and writing $2\sin\theta\cos\theta$ for $\sin 2\theta$, we obtain

$$\frac{\sin\theta}{n} = \frac{2\sin\theta\cos\theta}{n+2},$$

$$\cos\theta = \frac{n+2}{2n}.$$

Equating $\dfrac{n+2}{2n}$ to the expression of $\cos\theta$ obtained from the law of cosines yields

$$\frac{n+2}{2n} = \frac{(n+1)^2 + (n+2)^2 - n^2}{2(n+1)(n+2)}$$

$$= \frac{(n+1)(n+5)}{2(n+1)(n+2)} = \frac{n+5}{2(n+2)}.$$

Thus $n = 4$ and $\cos\theta = \dfrac{4+2}{4(2)} = \dfrac{3}{4}.$

24. (A) In the adjoining figure, let $AH = y$, $BD = a$, $DE = x$ and $EC = b$. We are given $AG = 2$, $GF = 13$, $HJ = 7$ and $FC = 1$. Thus the length of the side of the equilateral triangle is 16. Also, using the theorem about secants drawn to a circle from an external point, we have $y(y + 7) = 2(2 + 13)$, or

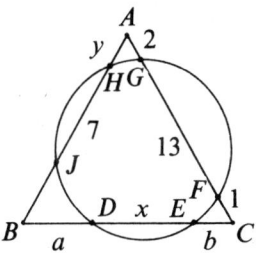

$$0 = y^2 + 7y - 30 = (y - 3)(y + 10).$$

Hence $y = 3$ and $BJ = 6$. Using the same theorem we have $b(b + x) = 1(1 + 13) = 14$ and $a(a + x) = 6(6 + 7) = 78$. Also, $a + b + x = 16$.
There are many ways of solving the system

(1) $a^2 + ax = 78,$

(2) $b^2 + bx = 14,$

(3) $a + b + x = 16.$

The one given below allows us to find x without first having to find a and b.

Subtract the second equation from the first, factor out $(a - b)$, and use (3):

$$a^2 - b^2 + (a - b)x = (a - b)(a + b + x)$$
$$= (a - b)16 = 78 - 14 = 64.$$

Therefore

$$a - b = 4.$$

Adding this to (3) we obtain $2a + x = 20$, whence

$$a = 10 - (x/2).$$

This, substituted into (1), yields

$$\left(10 - \frac{x}{2}\right)\left[10 - \frac{x}{2} + x\right] = \left(10 - \frac{x}{2}\right)\left(10 + \frac{x}{2}\right)$$

$$= 100 - \frac{x^2}{4} = 78;$$

$$\frac{x^2}{4} = 22, \qquad x = 2\sqrt{22}.$$

25. (D) The probability that the student passes through C is the sum from $i = 0$ to 3 of the probabilities that he enters intersection C_i in the adjoining figure and goes east. The number of paths from A to C_i is $\binom{2 + i}{2}$, because each such path has 2 eastward block segments and they can occur in any order. The probability of taking any

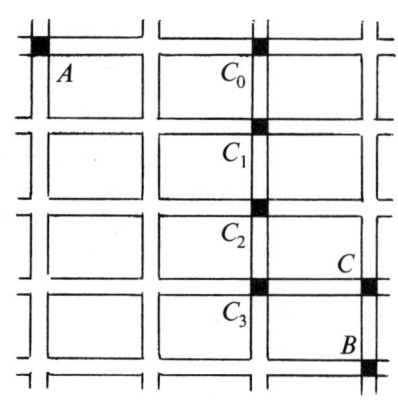

one of these paths to C_i and then going east is $\left(\frac{1}{2}\right)^{3+i}$ because there are $3 + i$ intersections along the way (including A and C_i) where an independent choice with probability $\frac{1}{2}$ is made. So the answer is

$$\sum_{i=0}^{3} \binom{2 + i}{2}\left(\frac{1}{2}\right)^{3+i} = \frac{1}{8} + \frac{3}{16} + \frac{6}{32} + \frac{10}{64} = \frac{21}{32}.$$

OR

One may construct a tree-diagram of the respective probabilities, obtaining the values step-by-step as shown in the scheme to the right (the final 1 also serves as a check on the computations).

It is important to recognize that not all twenty of the thirty-five paths leading from A to B through C are equally likely; hence answer (C) is incorrect!

$$1 \rightarrow \tfrac{1}{2} \rightarrow \tfrac{1}{4} \rightarrow \tfrac{1}{8}$$
$$\downarrow \quad \downarrow \quad \downarrow \quad \downarrow$$
$$\tfrac{1}{2} \rightarrow \tfrac{1}{2} \rightarrow \tfrac{3}{8} \rightarrow \tfrac{5}{16}$$
$$\downarrow \quad \downarrow \quad \downarrow \quad \downarrow$$
$$\tfrac{1}{4} \rightarrow \tfrac{3}{8} \rightarrow \tfrac{3}{8} \rightarrow \tfrac{1}{2}$$
$$\downarrow \quad \downarrow \quad \downarrow \quad \downarrow$$
$$\tfrac{1}{8} \rightarrow \tfrac{1}{4} \rightarrow \tfrac{5}{16} \rightarrow \tfrac{21}{32}$$
$$\downarrow \quad \downarrow \quad \downarrow \quad \downarrow$$
$$\tfrac{1}{16} \rightarrow \tfrac{3}{16} \rightarrow \tfrac{11}{32} \rightarrow 1$$

26. (B) If $n^2 = (ab3c)_8$, let $n = (de)_8$. Then $n^2 = (8d + e)^2 = 64d^2 + 8(2de) + e^2$. Thus, the 3 in $ab3c$ is the first digit (in base 8) of the sum of the eights digit of e^2 (in base 8) and the units digit of $(2de)$ (in base 8). The latter is even, so the former is odd. The entire table of base 8 representations of squares of base 8 digits appears below.

e	1	2	3	4	5	6	7
e^2	1	4	11	20	31	44	61

The eights digit of e^2 is odd only if e is 3 or 5; in either case c, which is the units digit of e^2, is 1. (In fact, there are three choices for n: $(33)_8$, $(73)_8$ and $(45)_8$. The squares are $(1331)_8$, $(6631)_8$ and $(2531)_8$, respectively.)

OR

We are given
$$n^2 = (ab3c)_8 = 8^3 a + 8^2 b + 8 \cdot 3 + c.$$

If n is even, n^2 is divisible by 4, and its remainder upon division by 8 is 0 or 4. If n is odd, say $n = 2k + 1$, then $n^2 = 4(k^2 + k) + 1$, and since $k^2 + k = k(k + 1)$ is always even, n^2 has remainder 1 upon division by 8. Thus in all cases, the only possible values of c are 0, 1 or 4. If $c = 0$, then $n^2 = 8(8K + 3)$, an impossibility since 8 is not a square. If $c = 4$, then $n^2 = 4(8L + 7)$ another impossibility since no odd squares have the form $8L + 7$. Thus $c = 1$.

27. (C) We recall the theorem that complex roots of polynomials with *real* coefficients come in conjugate pairs. Though not applicable to the given polynomial, that theorem is proved by a technique which we can use to work this problem too. Namely, conjugate both sides of the original equation
$$0 = c_4 z^4 + ic_3 z^3 + c_2 z^2 + ic_1 z + c_0,$$
obtaining
$$0 = c_4 \bar{z}^4 - ic_3 \bar{z}^3 + c_2 \bar{z}^2 - ic_1 \bar{z} + c_0$$
$$= c_4(-\bar{z})^4 + ic_3(-\bar{z})^3 + c_2(-\bar{z})^2 + ic_1(-\bar{z}) + c_0.$$

That is, $-\bar{z} = -a + ib$ is also a solution of the original equation. (One may check by example that neither $-a - bi$ nor $a - bi$ nor $b + ai$ need be a solution.) For instance, consider the equation $z^4 - iz^3 = 0$ and the solution $z = i$. Here $a = 0, b = 1$. Neither $-i$ nor 1 is a solution. [Alternatively, the substitution $z = iw$ into the given equation makes the coefficients real and the above quoted theorem applicable.]

28. (B) Let n be the last number on the board. Now the largest average possible is attained if 1 is erased; the average is then

$$\frac{2 + 3 + \cdots + n}{n - 1} = \frac{\dfrac{(n + 1)n}{2} - 1}{n - 1} = \frac{n + 2}{2}.$$

The smallest average possible is attained when n is erased; the average is then

$$\frac{n(n - 1)}{2(n - 1)} = \frac{n}{2}.$$

Thus

$$\frac{n}{2} \leqslant 35\frac{7}{17} \leqslant \frac{n + 2}{2},$$

$$n \leqslant 70\frac{14}{17} \leqslant n + 2,$$

$$68\frac{14}{17} \leqslant n \leqslant 70\frac{14}{17}.$$

Hence $n = 69$ or 70. Since $35\frac{7}{17}$ is the average of $(n - 1)$ integers, $(35\frac{7}{17})(n - 1)$ must be an integer and n is 69. If x is the number erased, then

$$\frac{\frac{1}{2}(69)(70) - x}{68} = 35\frac{7}{17}.$$

So

$$69 \cdot 35 - x = \left(35\frac{7}{17}\right)68$$
$$= 35 \cdot 68 + 28,$$
$$35 - x = 28,$$
$$x = 7.$$

29. (A) Let $m = x_0 y_0 z_0$ be the minimum value, and label the numbers so that $x_0 \leqslant y_0 \leqslant z_0$. In fact $z_0 = 2x_0$, for if $z_0 < 2x_0$, then by decreasing x_0 slightly, increasing z_0 by the same amount, and keeping y_0 fixed, we would get new values which still meet the constraints but which have a smaller product—contradiction! To show this contradiction formally, let $x_1 = x_0 - h$ and $z_1 = z_0 + h$, where $h > 0$ is so small that $z_1 \leqslant 2x_1$ also. Then x_1, y_0, z_1 also meet all the original constraints, and

$$x_1 y_0 z_1 = (x_0 - h)y_0(z_0 + h)$$
$$= x_0 y_0 z_0 + y_0\left[h(x_0 - z_0) - h^2\right] < x_0 y_0 z_0.$$

So $z_0 = 2x_0$, $y_0 = 1 - x_0 - z_0 = 1 - 3x_0$, and
$$m = 2x_0^2(1 - 3x_0).$$
Also, $x_0 \leqslant 1 - 3x_0 \leqslant 2x_0$, or equivalently, $\frac{1}{5} \leqslant x_0 \leqslant \frac{1}{4}$.
Thus m may be viewed as a value of the function
$$f(x) = 2x^2(1 - 3x)$$
on the domain $D = \{x | \frac{1}{5} \leqslant x \leqslant \frac{1}{4}\}$. In fact, m is the *small-est* value of f on D, because minimizing f on D is just a restricted version of the original problem: for each $x \in D$, setting $y = 1 - 3x$ and $z = 2x$ gives x, y, z meeting the original constraints, and makes $f(x) = xyz$.

To minimize f on D, first sketch f for all real x. (See Figure.) Since f has a relative minimum at $x = 0$ ($f(x)$ has the same sign as x^2 for $x < \frac{1}{3}$), and cubics have at most one relative minimum, the minimum of f on D must be at one of the endpoints. In fact,

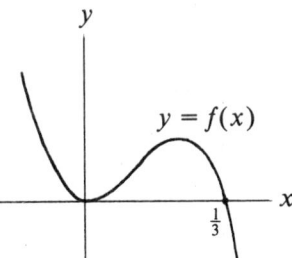

$$f\left(\tfrac{1}{4}\right) = \tfrac{1}{32} \leqslant f\left(\tfrac{1}{5}\right) = \tfrac{4}{125}.$$

30. (D) Let $d_1 = a + \sqrt{b}$ and $d_2 = a - \sqrt{b}$, where $a = 15$ and $b = 220$. Then using the binomial theorem, we may obtain
$$d_1^n + d_2^n = 2\left[a^n + \binom{n}{2}a^{n-2}b + \binom{n}{4}a^{n-4}b^2 + \ldots\right],$$
where n is any positive integer. Since fractional powers of b have been eliminated in this way, and since a and b are both divisible by 5, we may conclude that $d_1^n + d_2^n$ is divisible by 10.

We now apply the above result twice, taking $n = 19$ and $n = 82$. In this way we obtain
$$d_1^{19} + d_2^{19} = 10k_1 \text{ and } d_1^{82} + d_2^{82} = 10k_2,$$
where k_1 and k_2 are positive integers. Adding and rearranging these results gives
$$d_1^{19} + d_1^{82} = 10k - \left(d_2^{19} + d_2^{82}\right),$$
where $k = k_1 + k_2$. But
$$d_2 = 15 - \sqrt{220} = \frac{5}{15 + \sqrt{220}} < \frac{1}{3}.$$
Therefore, $d_2^{19} + d_2^{82} < 1$. It follows that the units digit of $10k - (d_2^{19} + d_2^{82})$ is 9.

Classification of Problems

To classify these problems is not a simple task; their content is so varied and their solution-possibilities so diverse that it is difficult to pigeonhole them into a few categories. Moreover, no matter which headings are selected, there are borderline cases that need cross-indexing. Nevertheless, the following may be helpful to the reader who wishes to select a particular category of problems.

The number preceding the semicolon refers to the last two digits of the examination year, and the numbers following the semicolon refer to the problems in that examination. For example, 82; 5 means Problem 5 in the 1982 examination.

Algebra

Absolute Value	73; 22 74; 11, 15, 27 75; 7
	77; 8 78; 9 79; 13 82; 19
Addition of Signed Numbers	78; 4 79; 4
Binary Operations	73; 5 74; 6 82; 7
Binomial Expansions (coefficients)	74; 3 75; 5 76; 23 77; 10
	82; 8, 30
Complex Numbers	74; 10, 17 76; 2 77; 16, 21
	80; 17 82; 27

Geometric	73; 28 74; 21, 28 75; 16 76; 4 77; 13 78; 24 80; 13 81; 14
Proportions	73; 8, 12, 29 74; 30 80; 10 82; 5
Radicals	75; 29 76; 29 79; 9 80; 27 81; 1, 29 82; 30
Reciprocals	76; 1 78; 2 81; 8
Remainder Theorem	74; 4 77; 28 79; 25 80; 24, 28 82; 1
Sequences (also see Progressions)	75; 15 79; 14, 7

Word Problems
  Age Problems
  Distance, Rate, Time
  General
 	
  Mixture
  Money
	
	76; 29 77; 12
	73; 27, 34
	75; 14, 17, 25 76; 13 78; 30
	80; 9 81; 4 82; 2
	73; 33 79; 15
	77; 3 78; 5

Arithmetic

Approximation	75; 29 76; 21 78; 18 79; 18
Arithmetic Mean	73; 2
Fractions	73; 13 75; 1 77; 24 79; 6 80; 1
Percent	73; 33 74; 7 81; 12, 13
Rationalization	74; 20 77; 7

Geometry

Analytic Geometry
  Circles
 	
  Lines
	
	73; 11, 30 78; 11 79; 8
	81; 27 82; 4
	75; 1, 2 80; 12, 22 81; 10
	82; 9

Trigonometry

Miscellaneous